FOCUS SERIES

Series Editor Alain Dollet

Thermochemical Conversion of Biomass for the Production of Energy and Chemicals

Anthony Dufour

ISTE

WILEY

First published 2016 in Great Britain and the United States by ISTE Ltd and John Wiley & Sons, Inc.

ISTE Ltd
27-37 St George's Road
London SW19 4EU
UK

www.iste.co.uk

John Wiley & Sons, Inc.
111 River Street
Hoboken, NJ 07030
USA

www.wiley.com

Library of Congress Control Number: 2015957826

British Library Cataloguing-in-Publication Data
A CIP record for this book is available from the British Library
ISSN 2051-2481 (Print)
ISSN 2051-249X (Online)
ISBN 978-1-84821-823-9

Contents

Acknowledgments

I sincerely thank my colleagues and friends who support me everyday in Nancy: Guillain Mauviel, Francis Billaud and Michel Mercy. I want to especially thank Gabriel Wild who often advised me and helped me edit this book.

I warmly thank the PhD, post-doctoral and Master's students I have (co-) supervised and who produced the majority of the results from our group presented in this book: L. Abdelouhaed, S. Cissé, X. Erdocia, J. François, T. Ghislain, L.Y. Jia, G. Lardier, Y. Le Brech, M. Nowakowska, R. Olcese, B. Ouartassi, M. Pander, C. Pelletier, L. Rodriguez, S. Rousseau, B. Shrestha and L. Tibavizco.

I also thank all the technicians and engineers from our laboratory and all the other colleagues with whom I have collaborated. I cannot mention everyone (more than 100 people) but be sure that I think of each of you when writing these lines.

Last but not least, I thank my family: my wife and my son. I hope he will see a beautiful, "smokeless", future.

Anthony DUFOUR
November 2015

Introduction on the Past, Present and Future of Biomass Conversion

I.1. Definition of biomass

"Biomass" is defined by the International Agency of Energy (IEA) as: "any organic, i.e. decomposing, matter derived from plants or animals available on a renewable basis. Biomass includes wood and agricultural crops, herbaceous and woody energy crops, municipal organic wastes as well as manure" [IEA 12].

In this book, we particularly focus on the "lignocellulosic" biomass, which is mainly wood or agricultural residue such as straw (Figure I.1). This type of biomass is today abundantly available on Earth and its valorization addresses important challenges for the development of sustainable energy and for mitigation of climatic changes.

Figure I.1. *This book focuses on the valorization of lignocellulosic biomass such as wood, bark, straw and pellets produced from wood sawdust*

This type of biomass is mainly composed of carbon (C), hydrogen (H) and oxygen (O). These atoms are distributed in biomass in complex macromolecules mainly composed of carbohydrates (cellulose and hemicelluloses) and aromatic compounds (lignin). Its chemical composition is described in detail in section 4.2.

I.2. Biomass in the global carbon balance of Earth

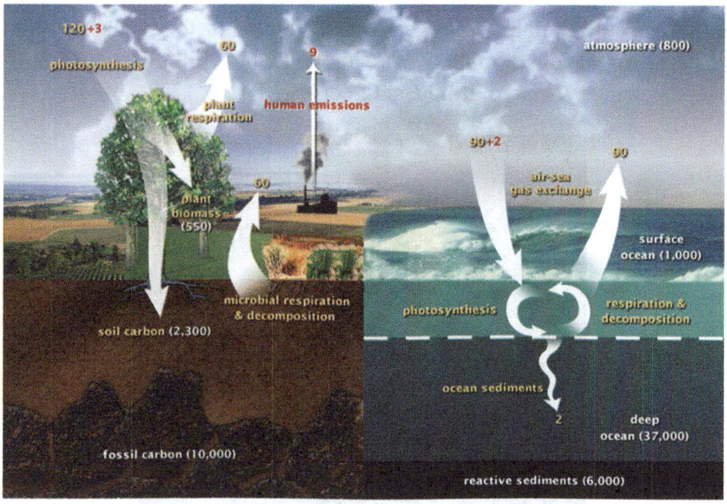

Figure I.2. *Diagram of the fast carbon cycle showing the movement of carbon between land, atmosphere and oceans [RIE 11]. Yellow numbers are natural fluxes and red are human contributions in gigatons (Gt) of carbon per year. White numbers indicate stored carbon (Gt). (Diagram adapted from U.S. DOE, Biological and Environmental Research Information System.) For a color version of the figure, see www.iste.co.uk/dufour/biomass.zip*

Carbon flows between each reservoir in an exchange called the carbon cycle, which has slow and fast components (Figure I.2). Any change in the cycle that shifts carbon out of one reservoir puts more carbon in the other reservoirs. Changes that put carbon gases into the atmosphere result in warmer temperatures on Earth. There are slow and fast carbon cycles. Over the long term (a few hundred thousand years), the slow carbon cycle seems to maintain a balance that helps keep Earth's temperature relatively stable. The fast carbon cycle (time measured in a lifespan) is largely the exchange of carbon through life forms on Earth. During photosynthesis, plants absorb carbon dioxide (CO_2) (120 + 3 gigatons of carbons/year (GtC/year)) to produce the biomass materials. This process is a part of the fast carbon cycle [RIE 11]. The emissions in the atmosphere due to the combustion of fossil carbon (9 GtC/year) affect the fast carbon cycle and result in an increase in the concentration of CO_2 in the atmosphere leading to the global warming of Earth. Theoretically, the current total consumption of fossil carbons (oil, coal and gas) would only represent the valorization of less than 10% carbon produced by biomass (10% of 120 GtC/year). Therefore, biomass could be (theoretically) a sufficient source of carbon to meet our energy needs. Hereafter we will see that the potential of biomass is in fact lower due to technical and economical constraints.

I.3. The history of biomass valorization

Wood has been the major source of fire for mankind for at least 600,000 years. In many parts of the developing world, more than 3 billion people still rely on biomass as the primary source of household energy [ANE 13].

Figure I.3. *More than 3 billion people rely on the use of biomass for cooking in inefficient open-fire stoves that particularly expose women and children to toxic indoor smoke [ANE 13]*

Figure I.4. *The traditional wood char production in developing countries*

Wood char has been produced for millennia for drawings (as in the cave of Grotte Chauvet, with drawings more than 38,000 years old), then for the manufacturing of bronze tools and as a cleaner alternative fuel than wood [ANA 03] (Figure I.4).

Figure I.5. *Hundreds of chemical commodities were produced from wood in the middle of the 19th Century. The process used was "wood distillation," which is now called "wood pyrolysis"*

Since the early 19th Century [MOL 08], wood "distillation" has been developed to produce char and also hundreds of chemicals (Figure I.5), such as acetic acid, methanol, acetone and formaldehyde, to produce various materials (e.g. Bakelite) and commodities. The industry of wood "distillation" or "carbonization" was extensively developed in the middle of the 20th Century with "advanced biorefineries" such as the plants located in the cities of Prémery and Clamecy in France. These important biorefineries valorized more than 100,000 tons of wood per unit (e.g. in Clamecy) [BRA 49]. This industry then declined with the development of fossil fuels (oil, coal and gas).

Wood was also used as a fuel for cars, especially during World War II. Wood (or char) was converted to a gas by a gasifier and then the gas was burnt in an engine. The first gasifier operated with wood char was developed by Cazes-Omnibus in 1910 and the gasifiers operated directly with wood by Georges Imbert in 1920.

The development of wood valorization for energy or chemicals by thermochemical processes is always related to the price or availability of fossil fuels. Figure I.6 describes the evolution of publications on wood gasification related to some international crisis or wars.

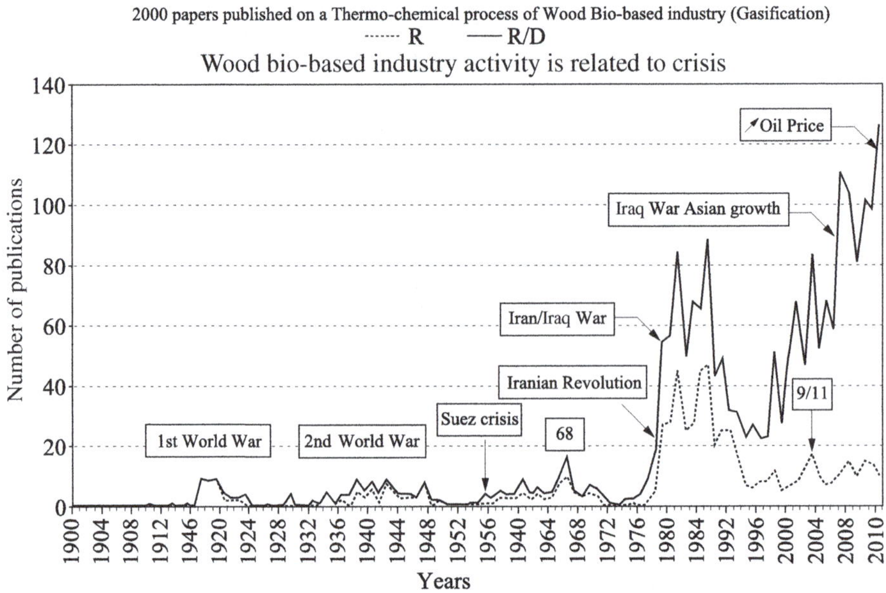

Figure I.6. *Evolution of the number of publications on wood gasification as a function of time highlighting the development of biomass conversion technologies after crisis (wars or increase in crude oil price) [DEL 08]*

More recently, the consumption of biomass has increased following the same global increase in the total world energy supply (Figure I.7) [IEA 14]. The global primary energy supply of biomass has increased mainly in developing countries (mainly in Asia and Africa) from 2000 to 2009 (Figure I.8). The share of biomass in the world total primary energy supply stays globally constant between 1973 (10.5% share) and 2012 (10.0%) (Figure I.7).

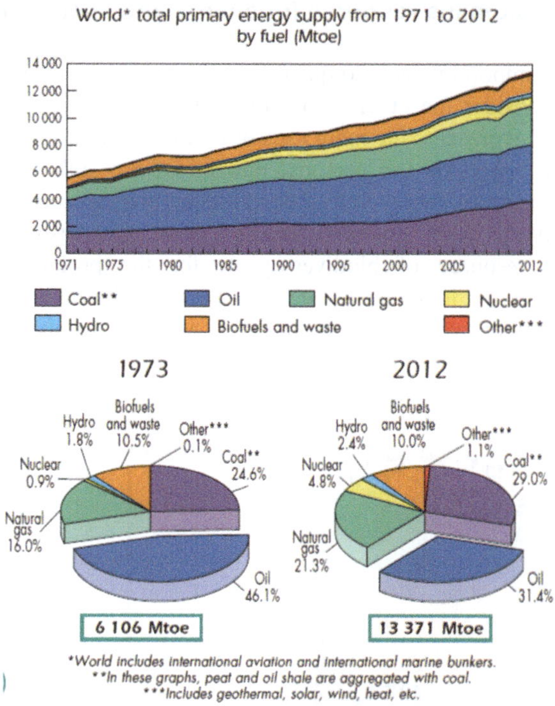

Figure I.7. *Evolution of the total primary energy supply in the world and share of biomass in 1973 and 2012 [IEA 14]. Mtoe means mega (10^6) tons oil equivalent. For a color version of the figure, see www.iste.co.uk/dufour/biomass.zip*

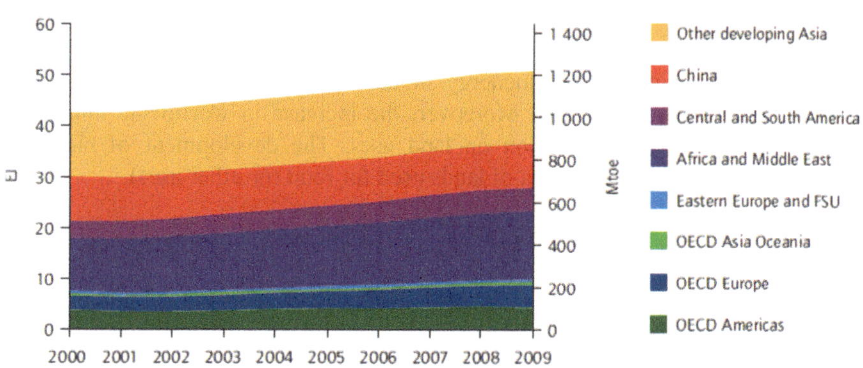

Figure I.8. *Evolution of the global (world) primary bioenergy supply [IEA 12]. For a color version of the figure, see www.iste.co.uk/dufour/biomass.zip*

I.4. Potential of biomass resource for energy and chemicals

The first very important issue to quantify the potential of biomass for energy or chemicals is to consider the other valorization of biomass, namely food, feed and materials. Figure I.9 presents the main uses of biomass with a rank on their priority use. This rank is based on the personal opinion of the author. This means that valorization of wood into energy is the last valorization to be considered if there is a competition with the other potential uses of wood. For instance, if the forest is an important resource of biodiversity or for the production of medicine (as biomolecules present in the leaves of trees), the trees should not be harvested and valorized for energy or even for timber. In other cases, wood should be first used for materials (timbers or paper) and then the wastes from these industries may be valorized for energy or bulk chemicals.

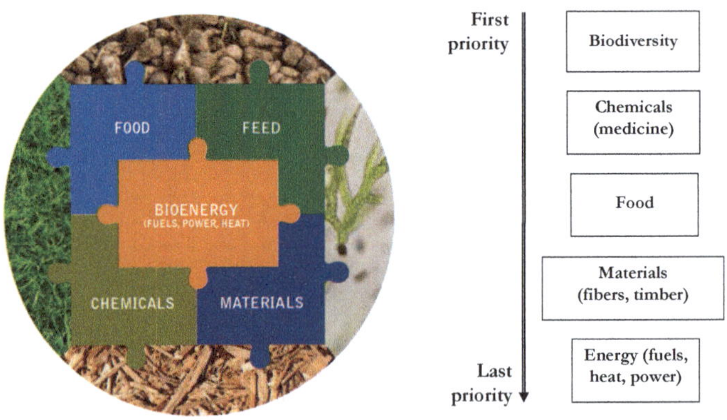

Figure I.9. *Various uses of biomass and priority use (personal opinion of the author)*

Therefore, the potential of bioenergy should be assessed by considering all the other potential uses of biomass. Moreover, the increase in worldwide inhabitants leads to an important competition on land uses. The development of bioenergy should not impact the surface area of lands used for food (or other uses).

Globally, the potential for bioenergy can be calculated as follows:

Land surface area available for energy = Total surface area – Surface area needed for food and other uses

Potential of biomass = Growth of biomass (e.g. expressed in m^3/ha/year) × Available surface area (ha)

The potential is absolutely not the total volume of wood standing on a piece of land (otherwise wood consumption will lead to deforestation) but the annual growth of biomass. Evaluating the available potential is a complex task: local environmental, technical, social and economical constraints have to be considered.

A comparison between the available potential and current use of biomass shows that on a worldwide level about two-fifths of the existing biomass potential for energy is used. In most areas of the world, the current biomass use is clearly below the available potential. The current use of biomass exceeds the available potential only in Asia [PAR 04].

The available potential for the world, Europe and France are presented in Table I.1.

Mtoe/year	Total primary energy supply	Current biomass consumption	Potential of biomass	References
World	12,000	1,200 (10%)	**2,200 (22%)**	[DES 92, PAR 04, IEA 14]
Europe	1,800	74 (4%)	**210 (12%)**	[DES 92, PAR 04, IEA 14]
France	270	14 (5%)	**26 (9%)**	[DUF 04]

Table I.1. *Potential of biomass for energy*

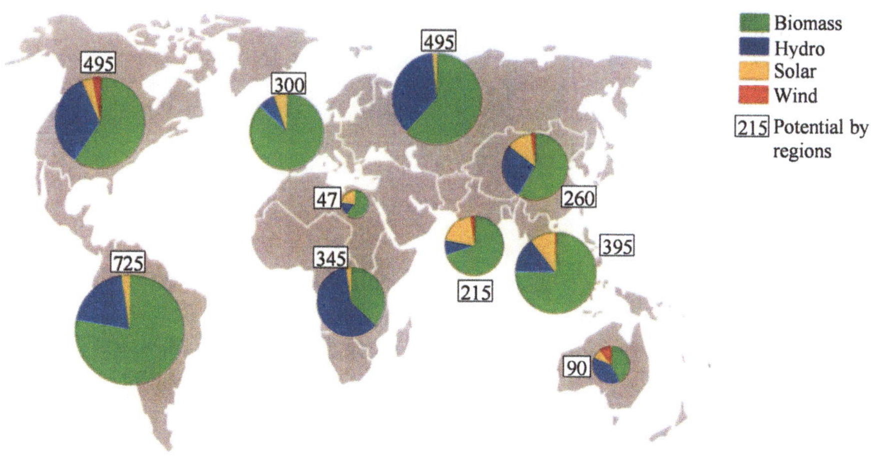

Figure I.10. *Potential of renewable energies in the world: biomass is the highest potential and well distributed (in toe/year) [DES 92]*

Globally, biomass could contribute to more than 20% of our energy needs if we were able to tackle the social, economical or political constraints. An interesting feature of biomass compared with fossil fuels is that it is well distributed over the world (Figure I.10) and would thus reduce the potential conflicts to control the energy resource (as for oil). Furthermore, biomass is not an intermittent resource as other renewable energies (wind and solar) are and it is a direct source of renewable carbon that could be converted into liquid biofuels.

More detailed roadmap visions for bioenergy development are proposed by the IEA. These studies are not detailed in this book [IEA 12].

I.5. Main routes for biomass conversion

I.5.1. *Issues to convert biomass into valuable products*

The first important issue to valorize biomass is its low-energy density, which results in high transportation costs. For this reason, small-scale (100 kWth) or medium-scale (< 10 MWth) units are preferred if biomass is not transformed into an energy vector with a higher energy density. Energy densities of various biomass fuels are presented in Figure I.11. Indeed, lignocellulosic biomass is a solid that is highly porous (low density) and composed of carbon, oxygen and hydrogen. The presence of oxygen reduces its calorific value compared to fossil fuels (which have very low oxygen contents).

The thermochemical processes aim at converting biomass into heat, electricity and liquid or gaseous biofuels. These energy vectors are then more convenient to be valorized than the solid biomass.

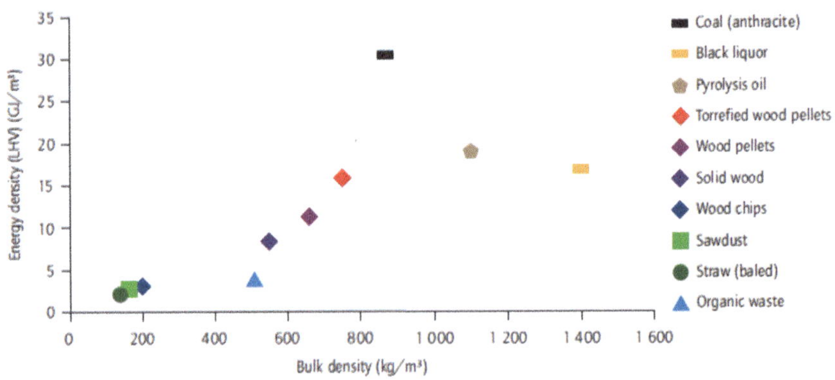

Source: IEA analysis based on DENA, 2011; FNR, 2011a; IEA Bioenergy, 2011; Kankkunen and Miikkulainen, 2003. For detailed data see Table 6 in Appendix I.

Figure I.11. *Energy density of various biomass fuels [IEA 12]. For a color version of the figure, see www.iste.co.uk/dufour/biomass.zip*

I.5.2. *Overview of thermochemical routes*

Figure I.12 displays the main thermochemical routes for biomass conversion.

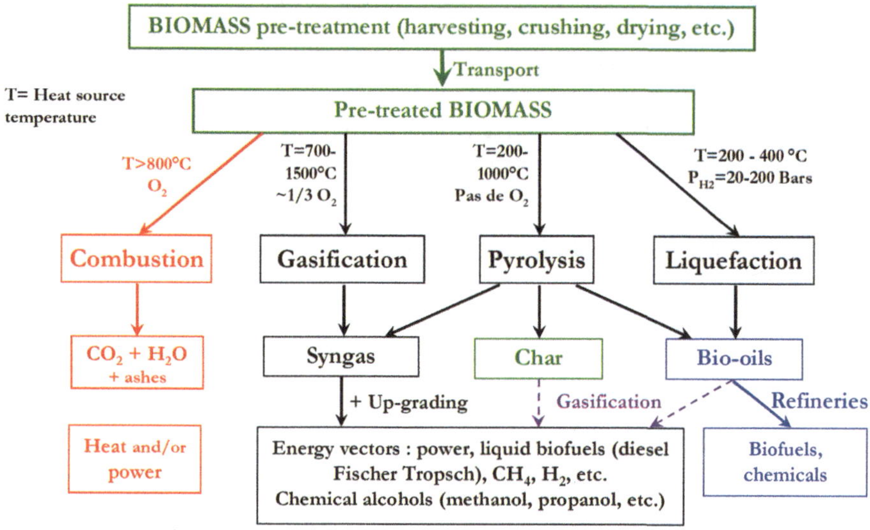

Figure I.12. *Overview of the main thermochemical processes to convert biomass into energy or chemicals*

From left to right, the processes are ranked as a function of the extent of oxidation of biomass: on the left, combustion (excess of air) to gasification (lower amount of air brought in the process) to liquefaction (reduction, no air).

The first and best known thermochemical conversion route is combustion. Wood combustion is carried out with excess air and at temperatures exceeding 1,000°C. The combustion allows a direct use of biomass to produce heat and/or electricity with a steam turbine.

The second route is biomass gasification. The gasification processes produce a syngas from biomass due to a controlled input of oxygen, which is roughly one-third of the amount of air required for combustion stoichiometry. The syngas is polluted with tar (by-products of gasification). Therefore, it should be cleaned (upgraded) to reduce its tar content for further valorization. The syngas could be used in gas engine to produce heat and power. It could be converted by catalysts to liquid biofuels (Fischer–Tropsch (FT) diesel or methanol) or to gaseous energy vectors (CH_4 or H_2).

A third route of biomass conversion is pyrolysis. Pyrolysis produces, by the action of heat and in an inert atmosphere, without any oxygen, a char, a bio-oil and a syngas. Char could be further used in gasification processes (such as biomass) or for higher added-value options (e.g. activated carbons). The bio-oil, which is a complex liquid of hundreds of oxygenated molecules, could be converted by special catalysts to biofuels or chemicals. These routes are detailed further in this book.

The liquefaction processes convert biomass directly into bio-oils, under high pressure of hydrogen, thus in a reducing atmosphere.

I.5.3. *Routes for bioelectricity production*

Figure I.13 presents the various technologies for electricity production from biomass. Electricity production should be combined with valorization of heat most of the time (combined heat and power (CHP) plant).

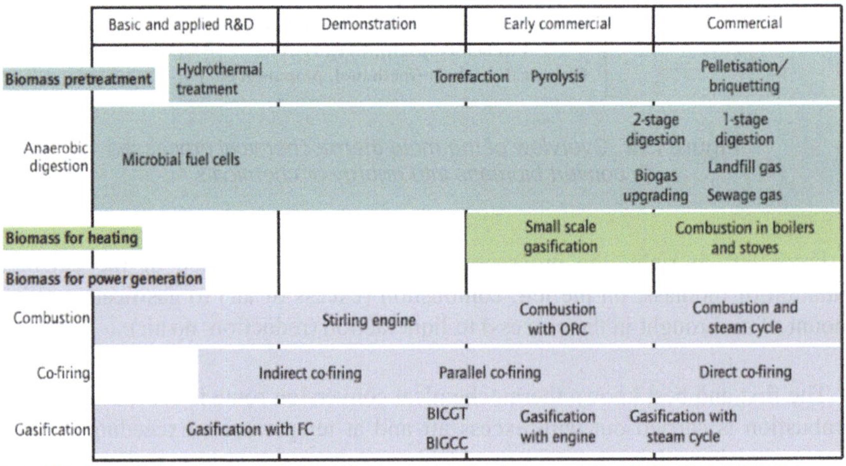

Note: ORC = Organic Rankine Cycle; FC = fuel cell; BICGT = biomass internal combustion gas turbine; BIGCC = biomass internal gasification combined cycle

Figure I.13. *Overview of technologies for electricity production from biomass and their current development status [IEA 12]. For a color version of the figure, see www.iste.co.uk/dufour/biomass.zip*

The main thermochemical technologies are combustion (combined with a steam cycle or turbine) and gasification (combined with gas engine or turbine). The direct combustion of pyrolysis bio-oil (not presented in Figure I.13) may also be an alternative technology.

I.5.3.1. *Technology 1: combustion followed by water–steam cycle*

Power production based on biomass combustion systems is an old and mature technology. Biomass is burnt in a boiler (e.g. stoker or fluidized bed) to produce high-pressure steam that is expanded totally or partially in a steam turbine. At different points of the process, residual thermal energy can be recovered for heat application, for example by extracting steam from the turbine.

I.5.3.2. *Technology 2: cofiring*

Biomass can be also cofired with coal in an existing coal-fired power plant, especially in large plants that cannot be applicable to dedicated biomass combustion because of limited local biomass availability. In direct cofiring, a percentage of biomass is added to the fossil fuel. Thus, it is possible to burn up to 5–10% of biomass on a heating value basis without extensive feedstock pretreatment and plant retrofitting. In parallel cofiring, biomass and coal are burnt separately in different boilers.

I.5.3.3. *Technology 3: gasification followed by syngas combustion*

Gasification systems convert biomass by partial oxidation at elevated temperatures to obtain a syngas mainly composed of CO, H_2, CO_2 and CH_4. The producer gas is then cooled and cleaned before being typically burnt in an internal combustion engine in the case of CHP decentralized production. Conditioning the producer gas (tar removal, etc.) to match the requirements of the gas engine is the main limitation for this technology.

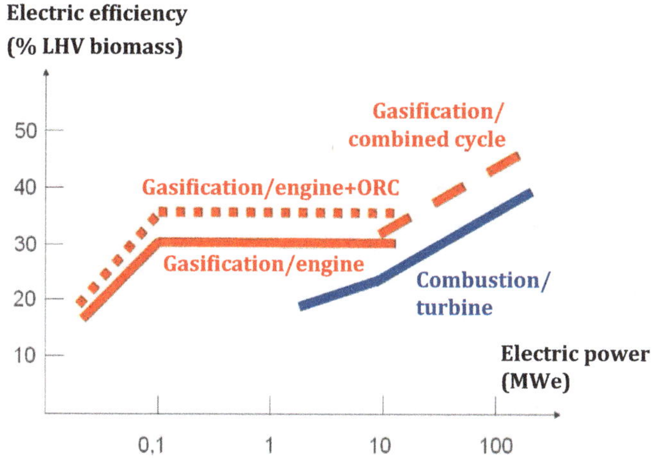

Figure I.14. *Electric efficiency of various technologies for electricity production as a function of the scale (LHV: lower heating value of biomass) [AUT 15]*

These technologies are adapted for different scales of electricity production (Figure I.14) and result in different efficiency. Gasification with a combined cycle (gas turbine and steam turbine) gives the highest electrical efficiency for large-scale units (>10 MWe).

I.5.4. *Examples of biorefinery concepts*

Figure I.15 presents the various consumer goods produced today from oil or gas. The petroleum-based refinery is based on some key commodity chemicals, such as benzene, xylene, ethylene and CO/H_2. Various catalytic processes produce the huge variety of finished products consumed today based on these few chemicals.

An analogous approach is developed for biorefineries. Biomass could be converted into various building blocks further converted by novel catalytic processes to biosourced products.

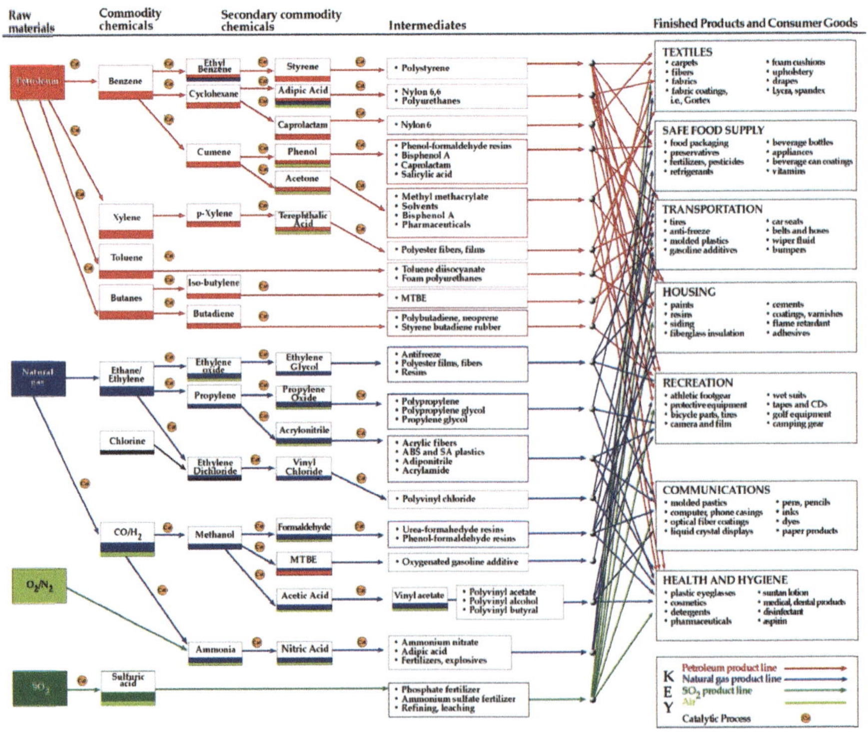

Figure I.15. *Flowchart for products from petroleum-based feedstocks [WER 04]. For a color version of the figure, see www.iste.co.uk/dufour/biomass.zip*

Figure I.16 presents a complete flowchart for the production of biobased goods from biomass.

The building blocks are ranked in Figure I.16 (from top to bottom) from light (H_2) to heavier molecules (C2 = molecules with 2 carbons to C6 = aromatics and macromolecules).

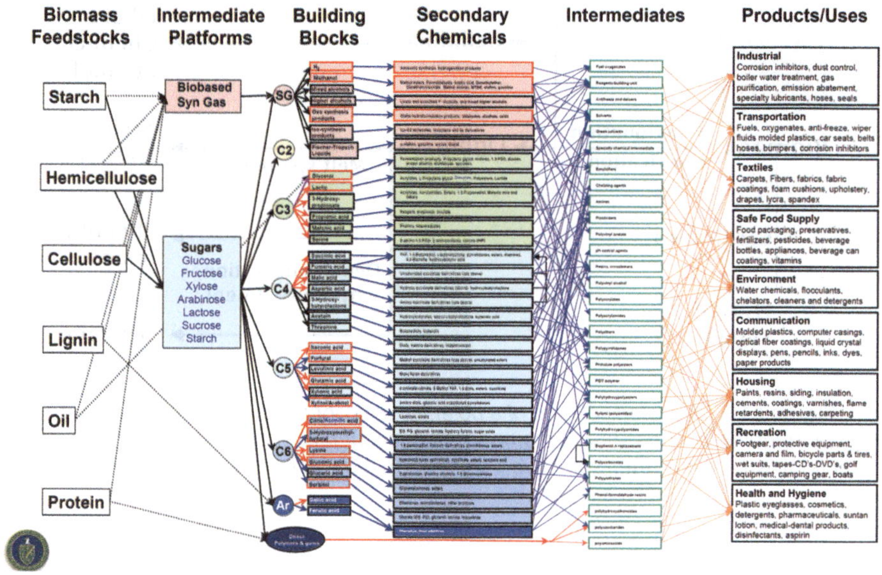

Figure I.16. *Analogous model of a biobased product flowchart for biomass feedstocks [WER 04]. For a color version of the figure, see www.iste.co.uk/dufour/biomass.zip*

Various concepts of biorefineries have been proposed to produce these various chemicals and fuels and have been classified depending on the feedstock and the process [CHE 09].

One option is to convert the whole wood in processes (for instance by gasification) to produce the building blocks. A second option is first to fractionate the macromolecules present in wood then to valorize them more selectively to produce chemicals, fuels and materials (Figure I.17). Wood is mainly composed of carbohydrates (cellulose and hemicelluloses) and of a phenolic macromolecule (lignin). The pulp and paper process is a first step toward "biorefinery" because this process separates the cellulose (for paper) from other polymers. It could be further modified to better valorize the by-products (hemicelluloses and lignin). For instance the Borregaard process is an advanced biorefinery currently operated at an industrial scale (Figure I.18).

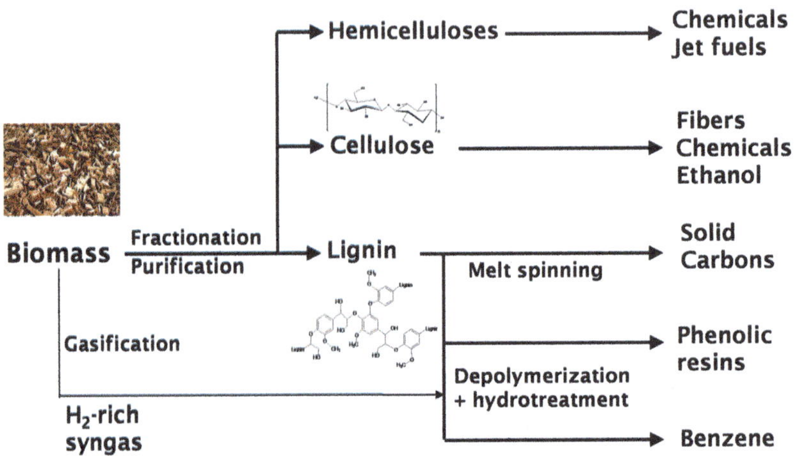

Figure I.17. *Simplified scheme of a potential concept of a biorefinery in which the macromolecules present in biomass (hemicelluloses, cellulose and lignin) are fractionated for a further selective valorization into fuels, chemicals and materials*

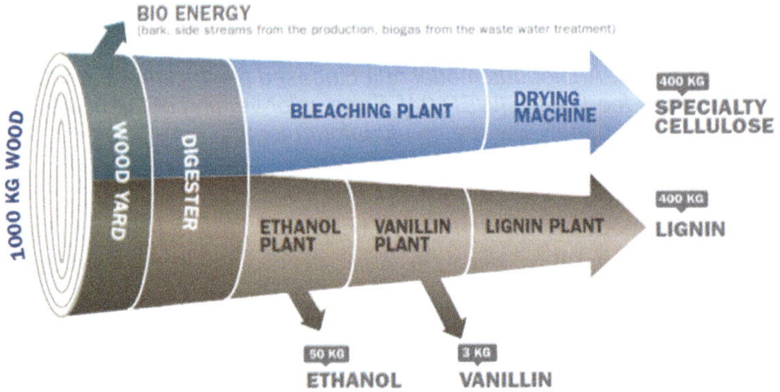

Figure I.18. *The Borregaard process: from specialty cellulose to biofuels and high added-value chemicals [RØD 12]*

Figure I.19 displays a complete scheme of the various biorefinery concepts that are classified based on biomass feedstocks and conversion processes.

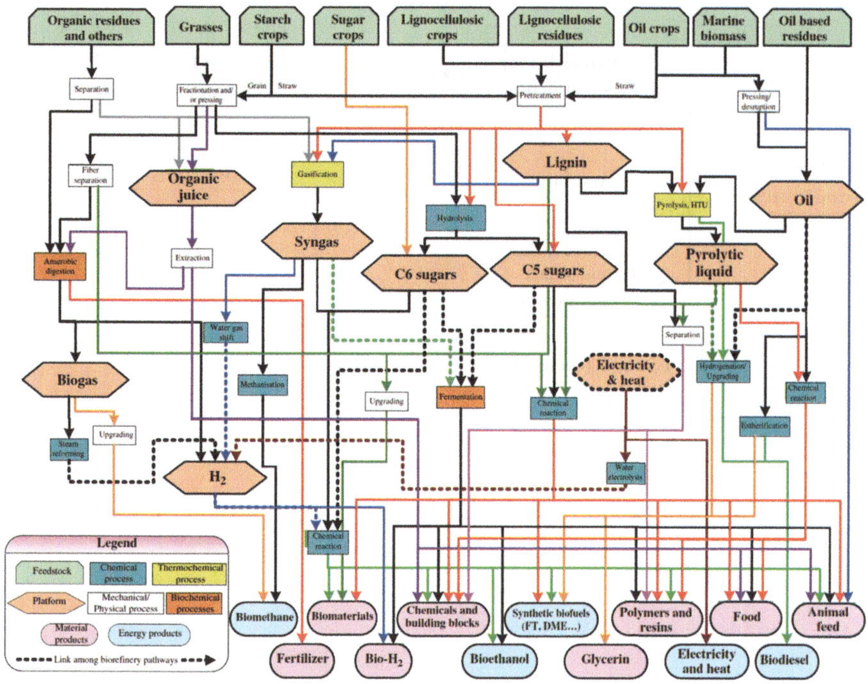

Figure I.19. *Network in which the individual biorefinery systems are combined [CHE 09]. For a color version of the figure, see www.iste.co.uk/dufour/biomass.zip*

I.5.5. *Routes for biofuels production*

In this section, we particularly focus on the main routes for biofuel production from lignocellulosic biomass. The main goal of the processes is to remove selectively (by catalysts) the oxygen present in biomass to produce liquid hydrocarbon-based biofuels that match the specifications of engines developed for petroleum-based fuels. The main routes for producing biofuels by thermochemical process on the whole biomass (without fractionation) are presented in Figure I.20. After pretreatment steps, biomass can be converted to syngas by gasification to produce the FT alkanes or bio-oils by fast pyrolysis or liquefaction. The bio-oil is then catalytically deoxygenated by specific catalysts. Bio-oils may be coprocessed with petroleum oil if the catalytic processes of the current refineries are adapted. The main interest of this second route is that the existing infrastructure of petroleum refineries is well suited for the production of biofuels. It may allow a more rapid transition to a sustainable economy without large capital investments for new reaction equipment (compared with the FT route) [HUB 07].

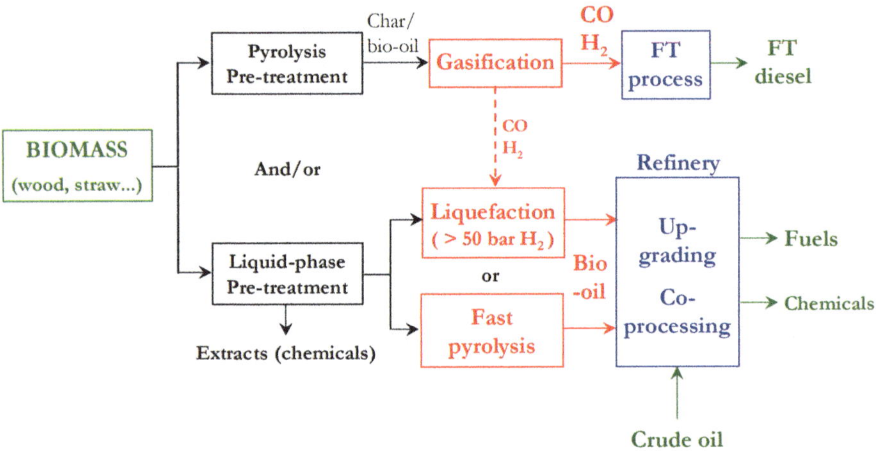

Figure I.20. *An example of routes for biofuel production from biomass thermochemical conversion*

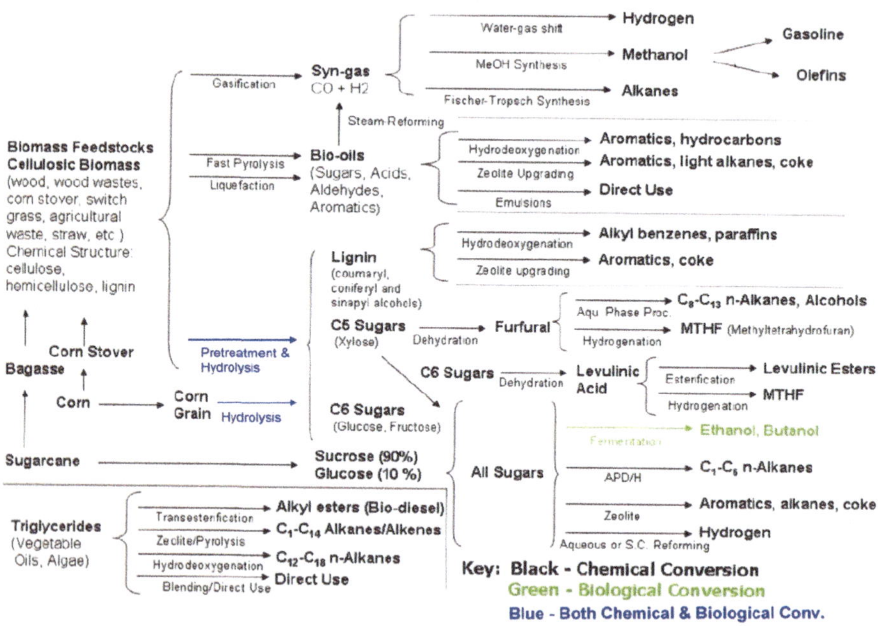

Figure I.21. *Comprehensive diagram of routes for biofuel production [HUB 08]. For a color version of the figure, see www.iste.co.uk/dufour/biomass.zip*

Figure I.21 presents a complete diagram of routes to produce biofuels. Both (thermo-)chemical and biological of biomass can be combined to produce biofuels. Hereafter, this book only focuses on thermochemical routes.

1

Bioenergy Chain and Process Scales

The outline of this book follows the different scales of investigation of biomass conversion which are presented in Figure 1.1. First the bioenergy routes are presented then followed by the reactor, particle and molecular scales.

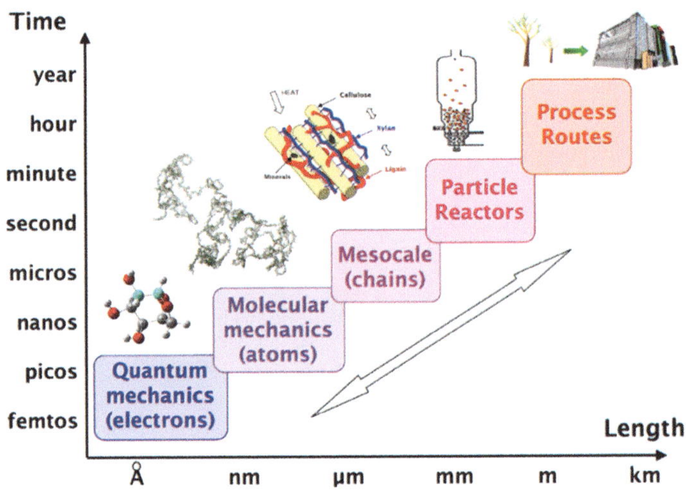

Figure 1.1. *The different scales of investigation in chemical engineering illustrated for biomass conversion: from electrons to routes*

1.1. Biomass production and ecological issues

There are two different contexts on biomass production:

– in some developing countries, biomass production may result in deforestation, a potential loss of biodiversity (especially in the rainforest), land-use competition and soil erosion (see Figure 1.2). Wood is often the main energy resource and is burnt in wood stoves with a very bad efficiency;

– in developed countries, forest is aging and in expansion due to the lack of wood valorization. Wood energy should be promoted in some regions to better valorize the small stems in forests.

Figure 1.2. *Two different contexts for wood valorization: (a) in developing countries (here in Madagascar), wood valorization results in deforestation and soil erosion; (b) in developed countries (as in France), wood is not well valorized and wood energy should be promoted to reduce the aging of trees and stems density in forests*

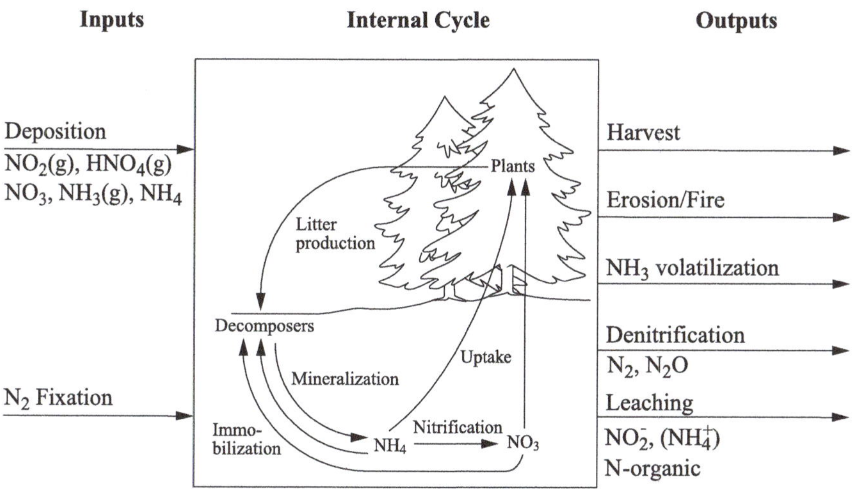

Figure 1.3. *Biogeochemical cycle of nitrogen in forests*

A very important feature for ecosystems is the biogeochemical cycle that is represented in Figure 1.3 for forests. The main inputs (especially in nutrients such as N, P and K) are atmospheric deposition and rock mineralization. The outputs are soil leaching or erosion and biomass harvesting. Wood harvesting can considerably impact the cycle if it results in too high a flow of nutrient exportation (nutrients are present in the harvested wood) and thus in soil impoverishment.

For these reasons, a global assessment of the whole bioenergy chain (from forest to final use) is required to assess the sustainability of bioenergy. It is especially of high importance to assess the fate of carbon and nutrients in the whole bioenergy chain.

1.2. Modeling of bioenergy chains

1.2.1. *Global model of the whole bioenergy chain*

Lignocellulosic biomass such as wood waste represents the highest potential of renewable resources but the biomass-to-energy route has to be complementary to other wood uses such as timber and pulp and should not yield to detrimental nutrient exportation issues. For this reason, our group has developed a modeling strategy (presented in Figure 1.4) that combines a forest modeling platform, called CAPSIS [DUF 12c, FOR 12], and a process modeling simulator (Aspen Plus®) [FRA 14].

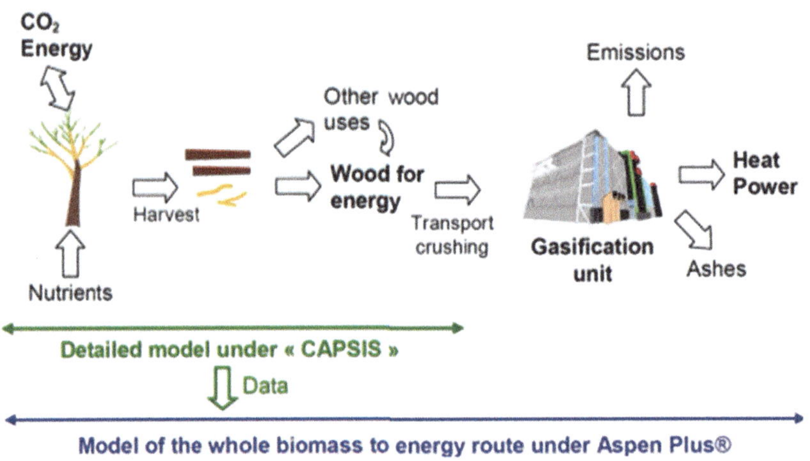

Figure 1.4. *Diagram of the approach developed at CNRS, Nancy, for bioenergy chain modeling: a forest management modeling tool ("CAPSIS") is combined with a process modeling simulator ("Aspen Plus") [FRA 14]*

CAPSIS predicts the biomass growth and CAT handles the different uses of the tree logs. The logs are distributed as a function of their diameter and quality to different wood valorization chains (pulp, timber, energy and end-uses). The results from CAPSIS and CAT are generated in an excel file in the form of "kg of biomass hectare^{-1}year^{-1}" including mineral flow rates (e.g. kg K hectare^{-1}year^{-1}). These results are included in an advanced Aspen Plus model. Aspen Plus is a piece of software that is well adapted to modeling solids, such as biomass, and their chemical conversion to liquids or gases. It yields mass and energy balances of the process. The forest growth and tree log chains have been included as "process units" in Aspen Plus. The forest is considered as a "reactor" that produces biomass from CO_2, sun power and other nutrients, based on the results of CAPSIS. This approach helps the simulation of the whole biomass to energy chains under a unique "process" flow sheet.

In Aspen Plus, the photosynthesis process is modeled in a reactor with the following simplified equation [FRA 14]:

$$aCO_2 + b/2\, H_2O + dN + eS + fCl + pP + kK \rightarrow C_aH_bO_cN_dS_eCl_fP_pK_k + (2a + b/2 - c)/2\ O_2$$

The flows in C, N, S, P and K (in kg ha^{-1}yr^{-1}) as estimated by CAPSIS were used as input parameters for the Aspen Plus photosynthesis reactors. The scope of this work was not yet to model in details the forest biogeochemical process but to propose a simplified method to handle forest management under a process flow sheet such as Aspen Plus.1

The forest growth process provides the amounts of energy wood by hectare and year. Given a particular amount of electricity and heat power, this integrated forest-to-energy model made it possible to predict the annual flows in wood, carbon and nutrients, including N, S, P and K, from the forest to the air emissions (NOx, SOx, polycyclic aromatic hydrocarbons (PAHs), etc.) and ash flows. The model of the gasification plant is described in the following section. On the basis of this approach, the required surface area of forest and other utilities (e.g. fuel for harvesting and transport) are linked with the consumption of the gasification unit. For instance, a need of 1 MW power yields a flow rate of wood chips (kg/h), calculated by the gasification module of Aspen Plus, and based on the forest growth module (kg/h/hectare) to a land use (hectare). Results from this integrated bioenergy chain model are presented in Figures 1.5 and 1.6.

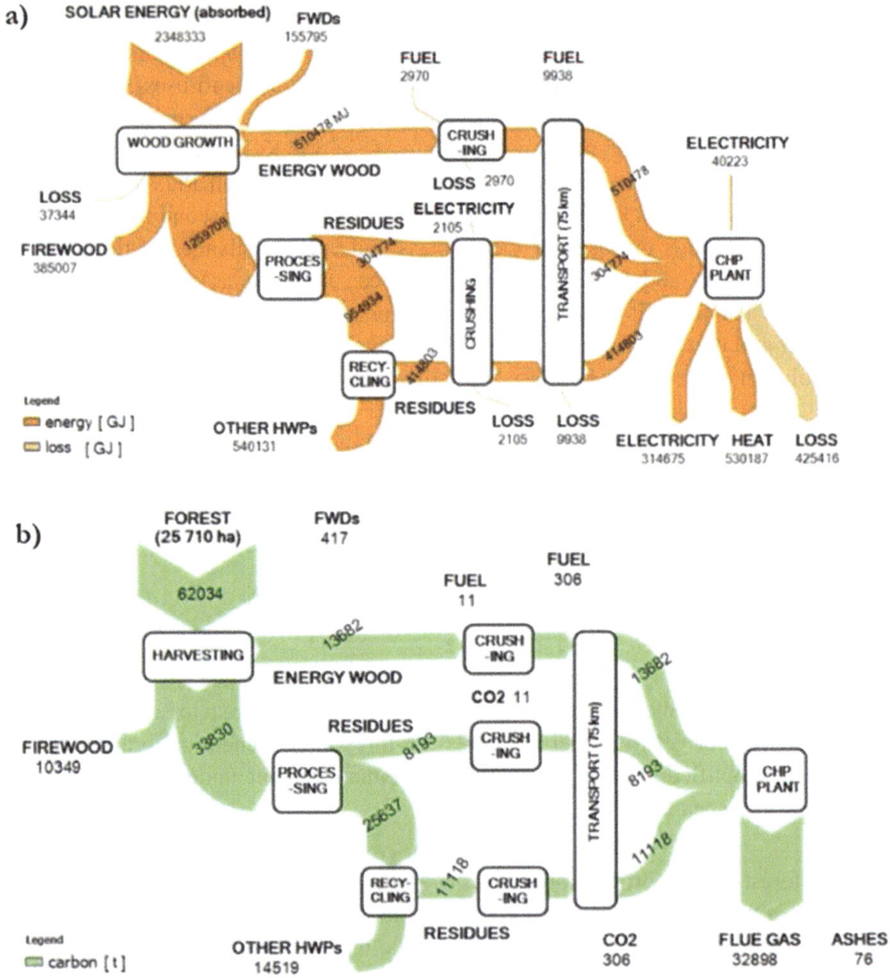

Figure 1.5. *(a) Energy and (b) carbon balance for the whole bioenergy chain from forest to power for the production of 10 MW of electricity during one year. "HWP" means "harvesting wood products"*

The bioenergy potential of pure even-aged high-forest stands of European beech, an abundant forest type in northeastern France, has been studied. Two forestry management practices were studied, a standard-rotation and a shorter-rotation scenario, along with two wood utilizations: with or without fine woody debris (FWD) harvesting (results not shown). FWD harvesting tended to reduce the forested area required to feed the combined heat and power (CHP) plant as larger amounts of energy wood were available for the CHP process, especially in the short-rotation scenario. The short-rotation scenario with FWD decreased the nutrient exports per hectare, compared to the standard practices, but increased the amount of nutrients in the CHP process. This increase in the input nutrient flows had direct consequences on the inorganic air emissions.

a)

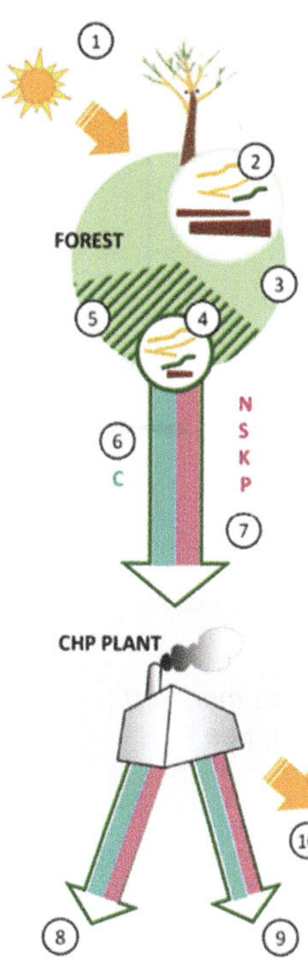

b)

		standard rotation		short rotation	
		without FWD	with FWD	without FWD	with FWD
1	Energy embedded in the total harvested biomass [TJ]	3 820	3 260	3 770	2 950
2	Total harvested biomass [Mg]	198 900	179 700	188 960	162 800
3	Total forested area [ha]	32 200	27 400	36 700	28 700
4	Total energy wood* to total harvested biomass [%]	35%	39%	37%	43%
5	Allocated forested area** [ha]	11 200	10 550	13 400	12 200
	Distribution of the energy wood products				
	Industrial energy wood [%]	57%	64%	63%	71%
	Industrial residues [%]	37%	31%	30%	23%
	Recycled HWPs [%]	6%	5%	7%	6%
	Flows of carbon and nutrients to the CHP plant				
6	C [Mg]	33 500	33 500	33 500	33 500
7	N [Mg]	65.1	67.4	66.3	69.6
	S [Mg]	7.97	8.08	8.00	8.15
	K [Mg]	88.6	87.2	76.0	76.6
	P [Mg]	6.53	6.86	6.64	7.10
	Flows of carbon and nutrients out of the CHP plant				
8	*Gaseous emissions*				
	C in CO_2 [Mg]	33 300	33 300	33 300	33 300
	C in CO [Mg]	105	105	105	105
	C in VOC [Mg]	44.7	44.7	44.7	44.7
	N in N2 [Mg]	36.1	38.2	37.2	40.2
	N in NO [Mg]	28.0	28.2	28.1	28.4
	N in N_2O [Mg]	0.737	0.766	0.750	0.791
	N in HCN [Mg]	0.223	0.232	0.227	0.240
	S in SO_2 [Mg]	6.77	6.86	6.79	6.92
	P in P_2O_5 [Mg]	0.0402	0.0426	0.0447	0.0473
	K in K_2O [Mg]	0.546	0.542	0.511	0.511
9	*Ash*				
	C [Mg]	55.0	55.0	55.0	55.0
	N [Mg]	0	0	0	0
	S [Mg]	1.20	1.22	1.21	1.23
	P in P_2O_5 [Mg]	6.49	6.82	6.60	7.05
	K in K_2O [Mg]	88.1	86.7	75.5	76.1
10	*Energy produced by the CHP plant*				
	Electricity [TJ]	281	281	281	281
	Heat [TJ]	567	567	567	567

*Total energy wood includes the industrial energy wood, industrial residues and recycled HWPs

** It is allocated by the mass ratio between total energy wood to total harvested biomass

Figure 1.6. *(a) Schematic diagram of the annual flows from the forest to the biomass CHP plant; (b) annual flows and area required to supply a CHP plant as a function of two forest management practices and two wood utilizations for the production of 10 MW electricity and 18 MW heat for one year [FRA 14]*

These mass and energy balances have since been used for a life-cycle assessment (LCA) of biomass gasification and combustion to produce heat and power.

1.2.2. *Model of a gasification plant*

Gasification of biomass could be a process of choice for producing heat and power with a higher efficiency of power production compared with combustion. Figure 1.7 presents the gasification CHP plant modeled under Aspen Plus and used in the model of the whole bioenergy chain [FRA 13a].

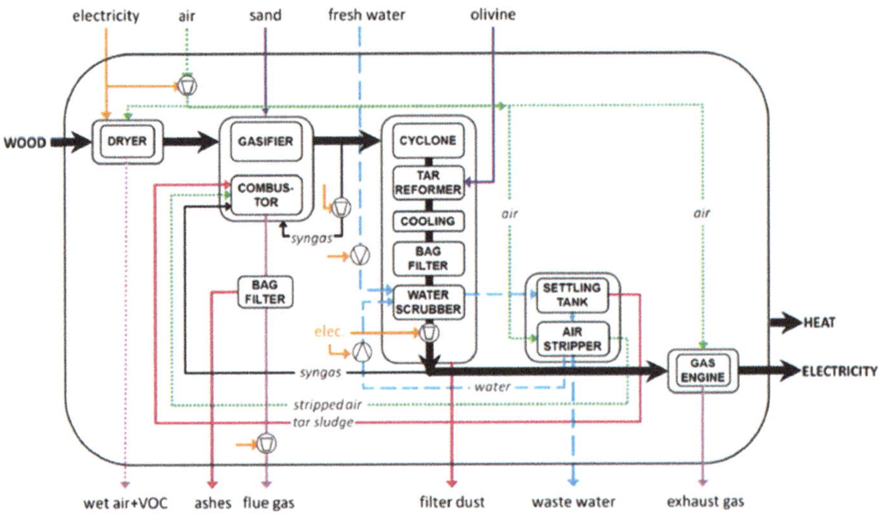

Figure 1.7. *Simplified diagram of the gasification process modeled under Aspen Plus. For a color version of the figure, see www.iste.co.uk/dufour/biomass.zip*

Schematically, the gasification CHP system consists of a wood dryer, an atmospheric dual-fluidized bed (DFB) gasifier, a syngas cleaning unit coupled with a scrubbing agent regeneration unit and a gas engine. Wood moisture content is reduced in the dryer using air as a drying agent. Dry wood is fed to the gasifier compartment of the DFB gasifier. A DFB gasifier consists of two interconnected fluidized bed reactors: one for gasification and one for char combustion. A sand bed is fluidized with a fraction of the raw syngas in the gasifier compartment and circulated in-between the two compartments to transfer heat from the combustor to the gasifier. In the gasification compartment, dry wood is converted into gas, char and tar. Char is injected with sand to the combustor compartment and oxidized with air. More details on the DFB model are given in section 2.3. Tar sludge (from the scrubbing agent regeneration step) and a fraction of clean syngas are additionally fueled into the combustor to maintain heat balance in the reactor. Flue gas and ashes exit the combustor zone whereas raw syngas exits the gasification zone.

Figure 1.8 shows the main compounds predicted by the model throughout the process.

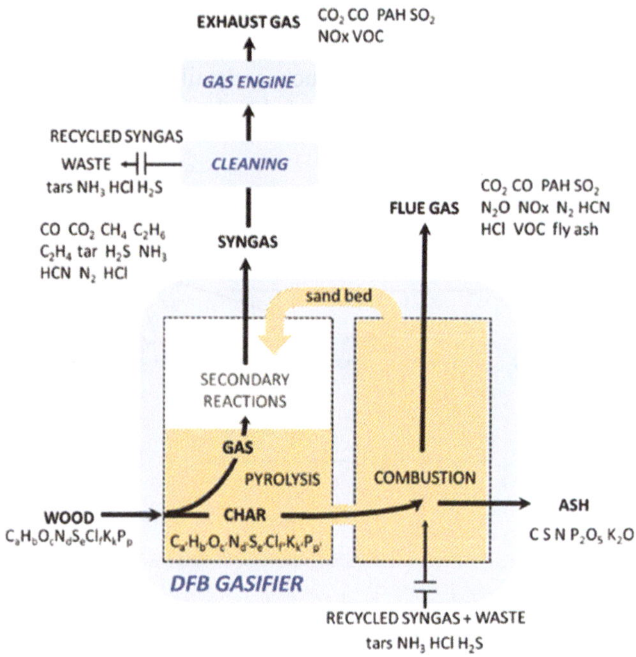

Figure 1.8. *Thermochemical conversion of the biomass and minerals throughout the gasification CHP process [FRA 14]*

Raw syngas contains various contaminants such as tar, NH_3, HCl and entrained particles, which are removed in a purification arrangement (Figure 1.8). The cyclone removes large particles; the tar catalytic reformer reduces the amount of tar and NH_3; the bag filter removes small particles and the water scrubber separates tar, NH_3 and HCl. Water is regenerated through a settling tank (for tar removal) and an air stripper (for NH_3 and HCl removal). In addition to reused water, freshwater is also continuously fed to the scrubber, and wastewater exits the regeneration unit. Stripped air is used as combustion air in the combustor. Clean syngas is then fed to the gas engine for power generation. Oil is consumed and exhaust gas emitted during gas engine operation.

An important feature of this Aspen Plus model is its suitability to model the evolution of trace/pollutant species due to a detailed description of the physical–chemical phenomena. For instance, N species contained in wood are converted to NH_3 and N moieties in the char in the gasifier. NH_3 in the syngas is further washed and oxidized to produce NOx (in the gas engine). N moieties in the char are oxidized in the combustor compartment of the gasifier and also generate NOx. Pollutant emissions from the engine and char combustor (NOx, SOx, PAHs, particles, Cl species, etc.) are predicted by the model based on the literature review. Tars (aromatic species) are also modeled. Such a detailed approach is realized by dedicated external Fortran files that are coupled to the Aspen Plus simulator.

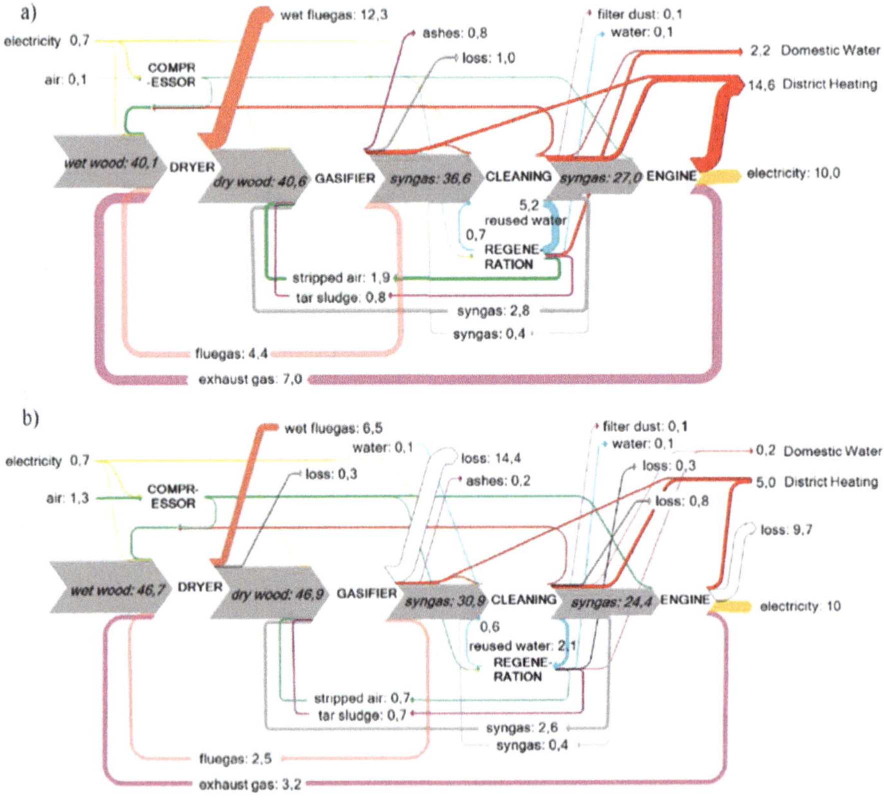

Figure 1.9. *(a) Energy and (b) exergy balances of the CHP gasification plant (streams in megawatts) [FRA 13b]. For a color version of the figure, see www.iste.co.uk/dufour/biomass.zip*

This detailed modeling approach has enabled us to calculate the mass, energy and exergy balances of different process configurations. A pinch analysis has been used to optimize heat recovery throughout the process. The influence of wood moisture content, drying process, flow rate of the sand circulating in the DFB reactor, catalyst type (olivine or Ni-based) for tar reforming and scrubbing agent efficiencies, as well as additional electricity production through steam turbine, on the overall process performances was reported [FRA 13b]. Figure 1.9 presents an example of Sankey diagrams derived from the model for a nominal process configuration.

The main exergy degradations occur in the gasification reactor and the engine (Figure 1.9(b)) where high-temperature reactions lead to important irreversibilities. The highest energetic efficiencies are obtained when no forced drying is implemented in the CHP system. Lowering the inlet wood moisture content with natural drying (energetically free) prior to the CHP plant improves the electrical efficiency. An overall energetic efficiency of 74% (23% electric and 51% thermal) is then reached with wood fed at 30% moisture content. The best exergetic efficiency is reached when wood (naturally dried to 30%) is additionally dried to 15% by forced drying in the CHP plant and when a part of the high-temperature heat is recovered for electricity production through steam turbine instead of being used for district heating. In this case, the overall energetic efficiency is 63% (32% electric and 31% thermal).

1.2.3. *Models of biorefinery sections*

To the best of our knowledge, there is still a lack of the literature on detailed modeling of the whole thermochemical biorefinery as it is presented for the biomass gasification chain. In the following sections, we present some examples of detailed modeling of biorefinery sections.

1.2.3.1. *Model of a partly integrated biorefinery: from biomass fractionation to lignin depolymerization*

The scope of this work is the production of aromatics from lignin liquefaction. The fractionation of biomass into carbohydrates and lignin has been experimentally studied by various methods [ERD 14] and then modeled into Aspen Plus. The quality and yield in lignins were compared. Then these lignins were converted into phenols by base-catalyzed depolymerization (NaOH in water under pressure). The work of CNRS-Nancy consists of the modeling and integration of all the streams (solids, liquids, solvents) and of closing the elemental balances (C, H, O) throughout the process (Figure 1.10). The main issue of all this work on modeling complex reactive systems is the closing of mass balance under Aspen Plus with "lumped pseudo-species" because an important part of compounds, notably heavy species, are not experimentally quantified.

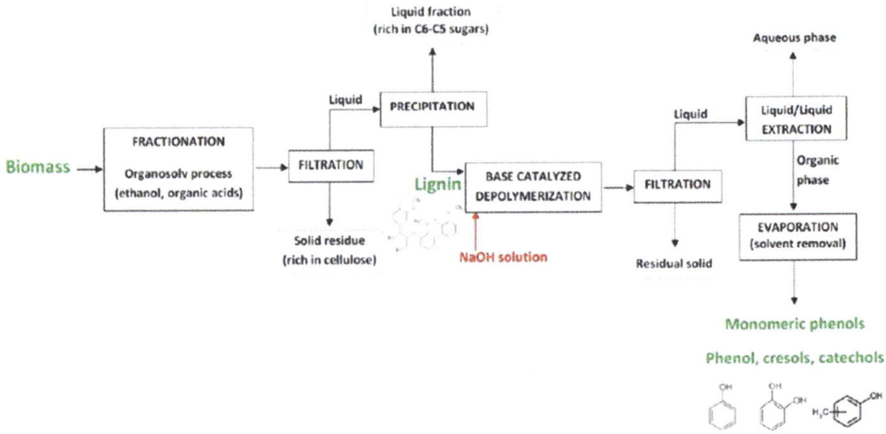

Figure 1.10. *Simplified scheme (without exchangers, pumps and solvents recycling) of the partly integrated process from biomass fractionation to lignin depolymerization*

1.2.3.2. *Model of aromatics production from lignin*

In one scenario of a future lignocellulosic biorefinery, cellulose and hemicelluloses would be depolymerized to be converted into ethanol and chemicals, thus generating considerable amount of lignin (~25 wt.% lignocellulosic biomass) [HOL 07, HAV 08].

Lignin is a very heterogeneous resource. Its composition and properties vary with biomass type, growing conditions and polymer fractionation process [HOL 07, ZAK 10]. However, lignin is always composed of aromatic monomers with different fractions in guaiacyl, syringyl or coumaryl groups [ZAK 10]. It is used to produce heat and power in the pulp and paper industry, but new valorization routes are needed to improve the revenue of lignocellulosic biorefinery [HUB 06, HOL 07]. It could be used to coproduce heat and power, a syngas, activated carbon [ROD 93], carbon fibers [KAD 02], aromatic hydrocarbons (benzene, toluene and xylenes (BTX)) [SHA 03, MA 12], fine chemicals (such as vanillin) [DEW 11, RØD 12], phenol [GOH 66], phenolic resins [EFF 08], etc.

Among these potential lignin uses, our research work has focused on the development of processes for the conversion of lignin into aromatic hydrocarbons

(Figure 1.11). The importance of this route for increasing revenues of biorefineries has been outlined in literature [SHA 03, HOL 07, HAV 08, ZAK 10].

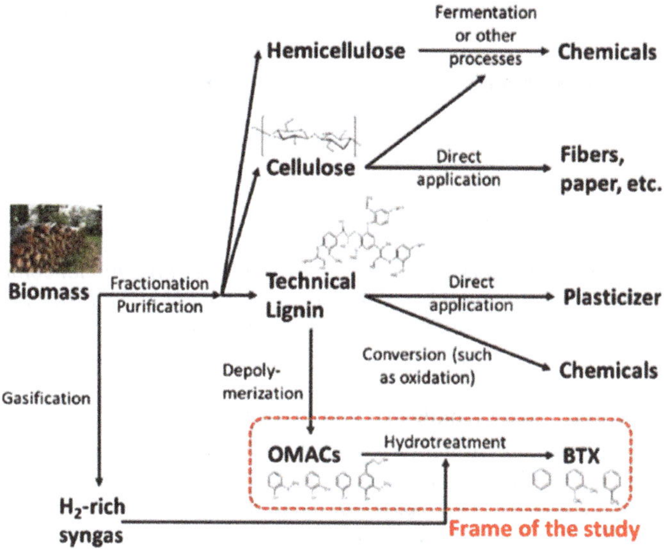

Figure 1.11. *A concept of biorefinery showing the frame of this study [OLC 12]: the production of aromatic compounds from lignin thermochemical conversion. "OMACs" means oxygenated monoaromatic compounds; "BTX" means benzene, toluene and xylenes*

Biosourced aromatics (BTX or phenols) could be produced by lignin pyrolysis coupled with catalytic hydrodeoxygenation (HDO) of uncondensed pyrolysis vapors. The vapor-phase hydrotreatment of pyrolysis vapors over cheap catalysts (iron-based catalysts) was the chosen strategy in this work instead of the liquid-phase hydrotreatment of bio-oils. Indeed, this process could reduce the problem of handling the reactive lignin oils and promote heat recovery. Moreover, the hydrogen partial pressure could be low (0.2 bar, typical of a syngas) and separating pyrolysis and catalytic reactors offer a more versatile process (different temperatures, regeneration cycles, etc.) than the direct catalytic pyrolysis. A second hydrotreatment refining step could be used on condensed bio-oils after the first HDO of vapors.

Guaiacol was used as a model compound to study the catalytic HDO over an Fe/SiO_2 catalyst. Experimental results were used to predict major and minor products by a semidetailed kinetic mechanism. The kinetic model is then included in an Aspen Plus model of lignin to BTX process (Figure 1.12).

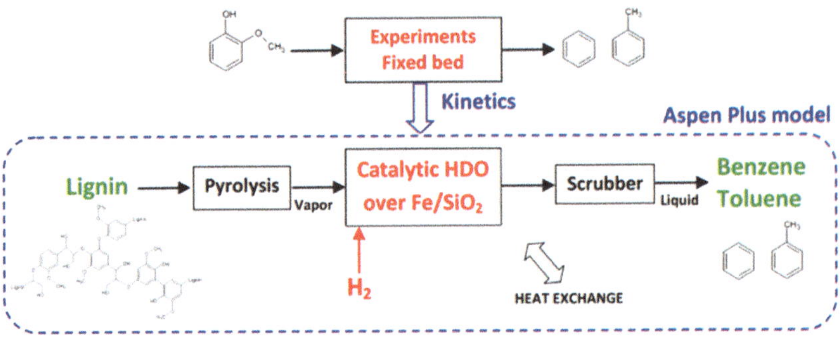

Figure 1.12. *Scope and methodology: experiments on a model compound (guaiacol) are used to model the whole lignin to BTX process under Aspen Plus [OLC 13]*

The Aspen Plus model handles (Figure 1.13): (1) pyrolysis of lignin, including char, oligomers, gases and aromatic yields; (2) catalytic conversion of aromatics by the kinetic model; (3) heat exchangers and (4) BTX vapor recovery by scrubbing with 1-methylnaphthalene. The elemental balances (C, H and O) of lignin pyrolysis were determined based on literature [DEW 09].

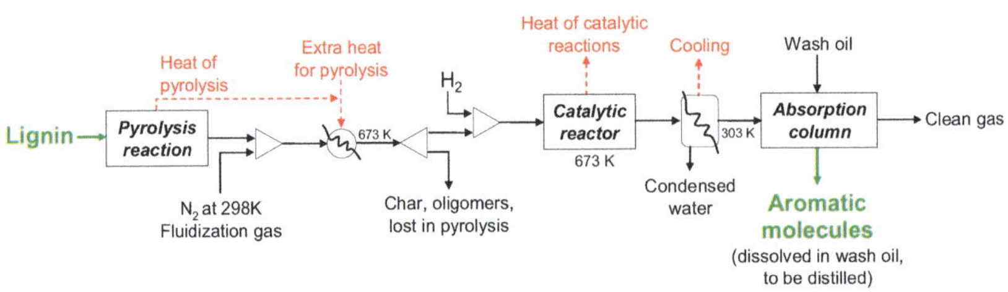

Figure 1.13. *Diagram of the lignin to BTX process by HDO of pyrolysis vapors and its simplification for modeling under Aspen Plus [OLC 13c]*

Figure 1.14 displays the mass flows obtained from Aspen Plus model (besides N_2 and 1-methylnaphthalene). It can be pointed out that the amount of catalytic coke is relatively low compared with that of char, oligomers and pyrolysis loss. Char, oligomers and pyrolysis loss lead to the major loss in mass and carbon.

The "gases" output in Figure 1.14 represents both non-condensable gases (CO_2, CH_4, H_2 and CO) and low amounts of condensable molecules (H_2O, benzene, etc.) that are entrained in the N_2 flow. Condensable products are mainly water (43.9 kg/h), benzene (17.8 kg/h), toluene (4.3 kg/h) and other compounds not converted by the catalytic reactor (650 kg), because the mass of catalyst was set only to match with a 0.1-kg/h phenol flow rate. If phenol becomes the desired product from lignin-derived pyrolytic vapors HDO over Fe/SiO_2, the mass of catalyst would be reduced from 650 kg (for 500 kg/h dry lignin and to reach 0.1 kg/h phenol if BTX are the desired products) to 52 kg to reach the maximum phenol yield.

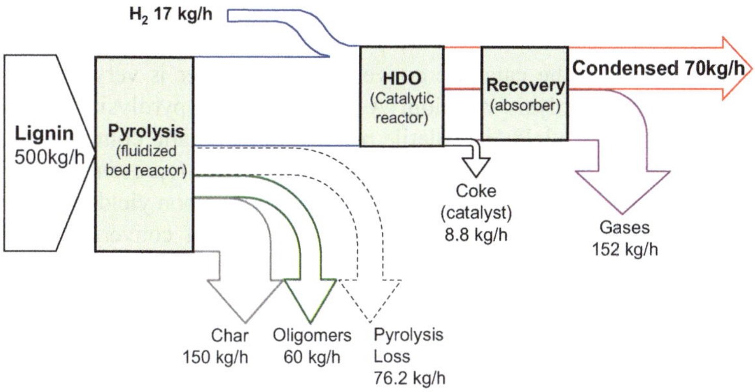

Figure 1.14. *Mass flows from Aspen Plus simulation of the lignin to BTX process. N2 (7,200 kg/h) and 1-methylnaphthalene (14.4 m3/h) (wash oil for recovery) are not represented. Condensed species (70 kg/h) are composed of 43.9 kg/h water, 17.8 kg/h benzene and 4.3 kg/h toluene (rest: methanol, cresols, etc.)*

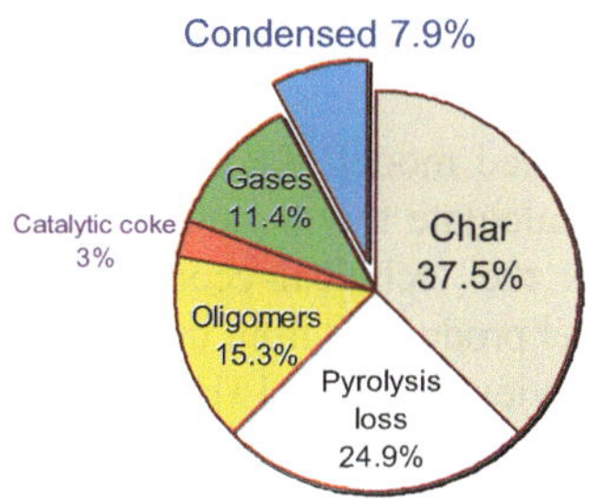

Figure 1.15. *Lignin-based carbon yield of lignin to BTX process by pyrolysis and gas-phase hydrotreatment at standard conditions (650 kg catalyst for 500 kg/h lignin, 7,200 kg/h N_2 carrier gas for a pyrolysis reactor). Condensed benzene and toluene account for 7.5% carbon*

Figure 1.15 shows the distribution of carbon yields in the different streams based on carbon in lignin. The char, oligomers and pyrolysis loss (carbon balance not closed during experiments) account for 77.7% carbon atoms from lignin, whereas the yield of benzene and toluene (BT) is only 7.5% carbon (4.4%wt. dry lignin). The species condensed by the scrubber, which are not aromatics (acetone, acetic acid), represent only 0.4% carbon. This yield in BT is still low and should be increased by process developments. The activity of the catalyst for the HDO of light aromatic compounds (lumped by guaiacol in the model) should even be lower in a real pyrolysis gas than that in the predicted gas (based on a model gas conversion) due to higher carbonaceous deposit. Consequently, the yield of desired products obtained from lignin experiments would most probably be lower than the predicted yield.

The carbon loss by the catalytic conversion or scrubber is very low compared with that from the pyrolysis reaction. A more efficient pyrolysis technology is needed to obtain higher yields in volatile oxygenated aromatic compounds that can be further converted into chemicals. The pyrolysis loss represents a substantial amount of missing carbon, three times (24.9% on lignin carbon yield) as much as the desired product (7.5%). An important loss in carbon, not converted into desired products, can also be attributed to oligomers (15.3%). However, solid carbons from pyrolysis can be valorized for heat production in the biorefinery or for higher added-value products.

The effect of gas dilution from a pyrolysis reactor on BTX losses, heat demand and scrubbing solvent flow rate has also been highlighted in this study. We have shown in this study [OLC 13c] that high carrier gas flow rates (as required for biomass pyrolysis in fluidized bed) lead to the entrainment of fines and oligomers, dilute the products and impact considerably the process.

1.3. Technical–economical analysis of bioenergy chains

The development of detailed model of bioenergy chains (e.g. from Aspen Plus simulations) is required to calculate the mass and energy balances of the processes and then to further assess the technical–economical feasibility. The technical–economical analysis of biofuel production from thermochemical conversion of biomass has been comprehensively reviewed [BRO 15] and is not presented in this book.

1.3.1. *Technical–economical analysis of power production*

The cost of electricity production from biomass decreases considerably with the scale of the unit (Table 1.1).

A very detailed study has been performed by Bridgwater *et al.* [BRI 12] on the comparison of the cost of electricity production from four technologies: pyrolysis followed by combustion of bio-oils in engine, gasification followed by gas engine or combined cycle (IGCC) and combustion associated with a steam cycle.

Capacity	< 10 MW	10–50 MW	>50 MW	Co-firing*
Typical power generation efficiency (%)	14–18	18–33	28–40	35–39
Capital costs (USD/kW)	6,000–9,800	3,900–5,800	2,400–4,200	300–700
Operating costs (% of capital costs)	5.5–6.5	5–6	3–5	2.5–3.5

* Co-firing costs related only to the investment in additional systems needed for handling the biomass fuels, with no contribution to the costs of the coal-fired plant itself. Effeciences refer to a plant without CCS.

Source: IEA analysis baed on DECC (2011), IPCC (2011), Mott MacDonald (2011), Uslu *et al.* (2012).

Table 1.1. *Overview of bioenergy power plant conversion efficiencies and cost components [IEA 12]*

The cost of electricity is between 0.2 and 0.25 €/kWhe at 1 MWe (electricity) and decreases to 0.1 €/kWhe at 20 MWe (Figure 1.16). The fast pyrolysis and diesel engine system is handicapped by the typical characteristics of all novel technologies: high capital costs, high labor and low reliability. The more established technology (combustion and steam cycle) produces lower cost electricity under current conditions. The fast pyrolysis and diesel engine system have relatively low system efficiency at high capacities because of a low conversion efficiency of biomass into pyrolysis liquids and char by-product.

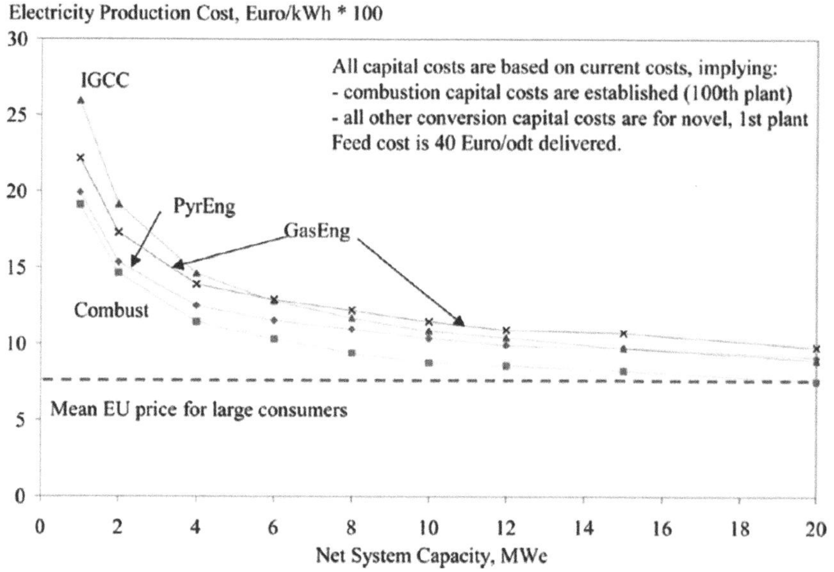

Figure 1.16. *Comparison of electricity production costs for four biomass-to-electricity systems [BRI 02]*

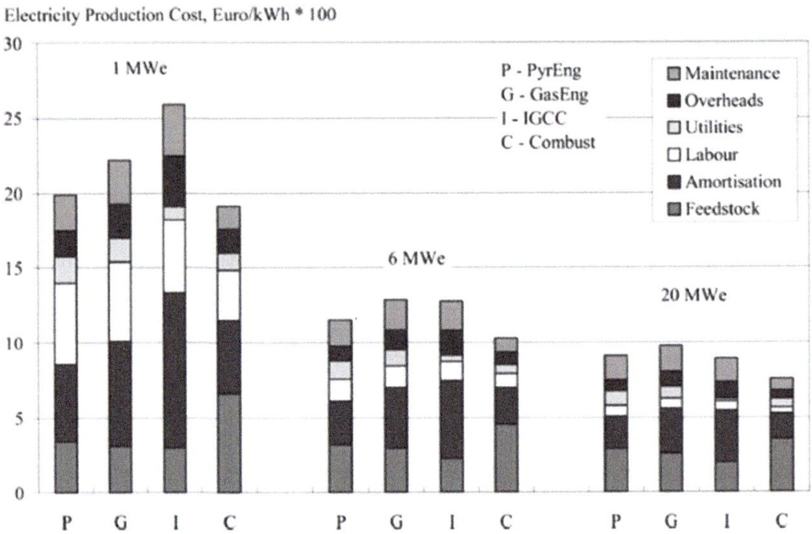

Figure 1.17. *Breakdowns of electricity production costs by cost sector for four biomass-to-electricity systems [BRI 02]*

The distribution of electricity cost by sectors is presented in Figure 1.17. The main costs are related to feedstock and amortization for all technologies. The capital costs decrease very significantly for all technologies with the scale of the units (Figure 1.18).

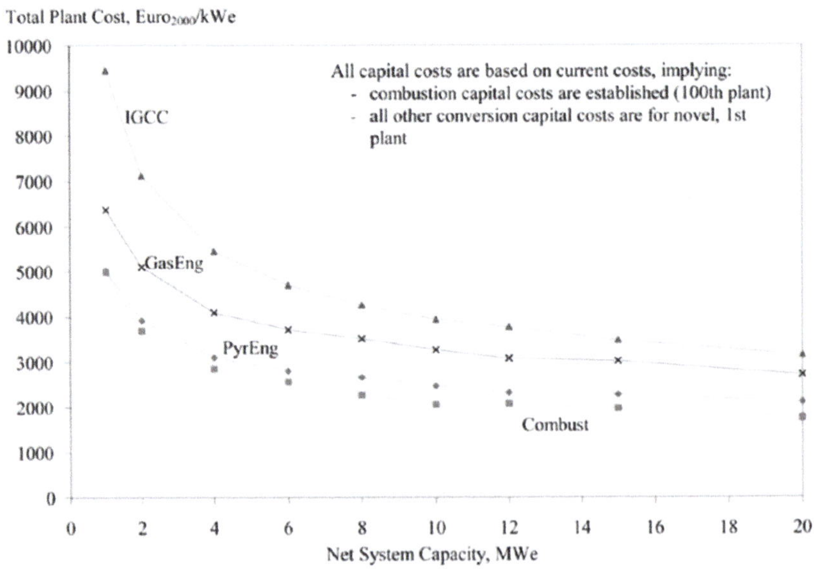

Figure 1.18. *Effect of system capacity on the total plant cost [BRI 02]*

1.3.2. *Technical–economical analysis of heat production*

Our group has undertaken a study [PEL 15] on the technical–economical analysis of heat production from biomass combustion (Figure 1.19) in collaboration with EIFER (Karlsruhe, Germany) and CIRED (Nogent, France) laboratories.

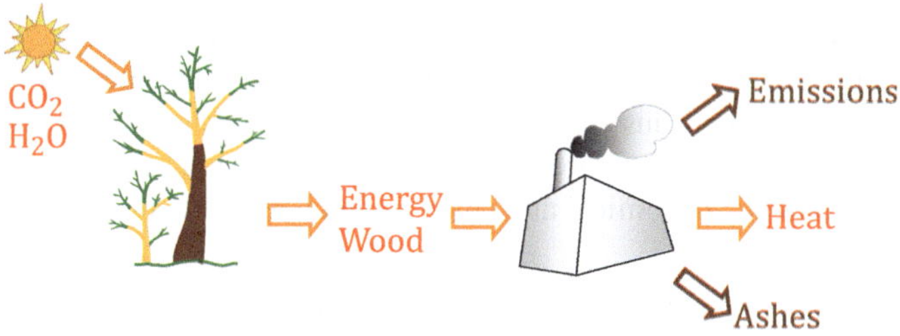

Figure 1.19. *System boundary of the study (from forest to wood pretreatment, combustion and emissions, and final heat). This whole bioenergy route is modeled under Aspen Plus software for a detailed calculation of mass and energy balances. For a color version of the figure, see www.iste.co.uk/dufour/biomass.zip*

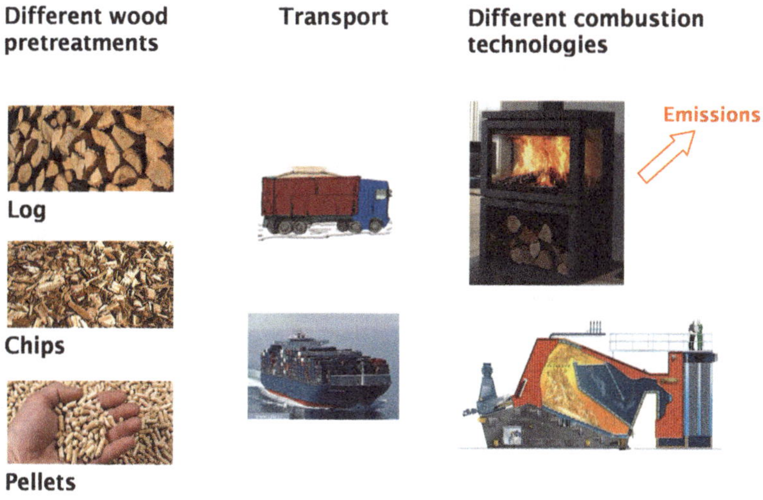

Figure 1.20. *Different options for wood combustion and heat production studied in a technical–economical analysis*

The scope of this study is to assess the technical–economical feasibility for the heating of a 100 m^2 house in Lorraine (France), with heating needs of 6,500 kWh/year and a hot water production for a family of four people of 2,500 kWh/year. Various technologies and process options are studied (Figure 1.20). These bioenergy routes are compared with conventional electric heating (with the French electricity mix) or gas boiler.

Our approach consists of a detailed modeling of the whole bioenergy chain from forest to combustion devices under Aspen Plus (as presented previously for gasification). The Aspen Plus model gives the complete mass and energy balances of the bioenergy routes (from forest growth to final heat). These balances are then used for the technical–economical and environmental assessments.

Table 1.2 gives the main data used for the technical–economical analysis and for the various technologies. Table 1.3 displays the costs and emissions of fuels.

	Heating/ hot water coverage (%)	**Thermal yield**	**Electrical needs of equipment (kWh/year)**	**Total investment cost (equipment + labor) (€)**	**Available tax cuts (€)**	**Maintenance cost (€/year)**
Log stove	50/0	Calculated by model	-	1,500 + 300	450	200
Log boiler	100/100	Calculated by model	400	6,500 + 1,500	1,950	250
Pellet stove	70/100	Calculated by model	200	2,800 + 500	750	200
Pellet boiler	100/100	Calculated by model	600	13,500 + 1,500	4,050	250
Condensing gas boiler	100/100	105%	30	5,000	-	250
Electric heating	100/100	100%	-	1,000 + 200	-	-
Wood district heating	100/100	Calculated by model, with heat losses in network	16	4,000	-	80
Gas district heating	100/100	85% including heat losses	200	2,000	-	50

Table 1.2. *Technical–economical parameters for combustion equipment [PEL 15]*

Fuel	Price for an individual user	Price for a large facility (district heating scenarios)	Impact of production and transport of solid fuels[a] (kg CO_2eq/ton)	Global impact of use
Wood logs (50 cm)	170 €/ton	N/A	20	Dependent on technology and combustion conditions, from the modeling under Aspen Plus
Wood pellets	280 €/ton	N/A	155	
Wood chips	N/A	80 €/ ton[b]	61[b]	
Natural gas	6.6 c€/kWh_{LHV}	5.1 c€/kWh_{LHV}	-	0.234 kg CO_2eq/kWh_{LHV}
Electricity (French mix)	15.1 c€/kWh	13.8 c€/kWh	-	0.18 kg CO_2eq/kWh_{LHV}

[a]For fossil fuels and electricity, the impact of production and transport is included in the global value.
[b]For a transport range of 150 km.

Table 1.3. *Prices and impact of production of wood fuels*

1.4. Life-cycle and multicriteria assessment of bioenergy chains

1.4.1. *Methodology of life-cycle assessment applied to bioenergy routes*

The LCA methodology is used to assess the processes or products from an environmental point of view.

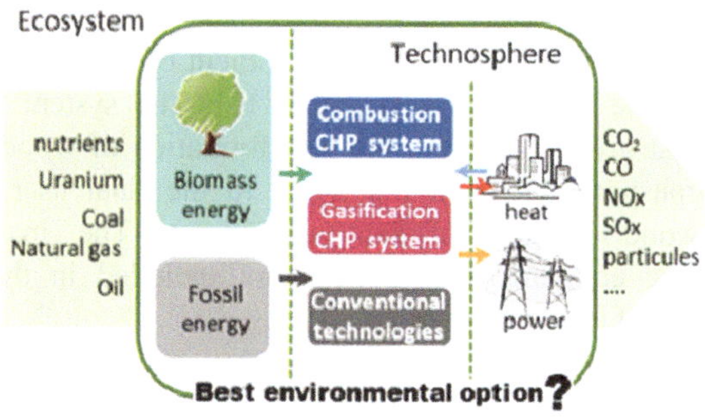

Figure 1.21. *From the ecosystem to the technosphere. The technosphere (processes, products, etc.) impacts the ecosystem*

Figure 1.21 describes the general methodology of LCA applied to the production of heat and power. The goal of LCA is to define what the "best process" is from an environmental point of view.

The first step in the LCA methodology is the definition of the function and functional units.

For instance, a function of the process can be to produce electricity and heat for a district heating or to produce a biofuel or to produce only heat in a house.

Then the functional units should be defined. It could be, for instance, the production of 1 kWh heat in a house or the production of 1 kWh heat for district heating and 0.4 kWh electricity. It could also be the valorization of 1 kg wood to compare different processes giving different products. The functional unit could also be 1 hectare of land to compare various uses of the land or even one defined territory.

Then the system boundaries should be defined. For instance, the life cycles of the biomass plants can be from cradle to grave: from forest growth to energy production. It could be only from dry wood input of the plant (without the transport) to the final product (heat or biofuels or chemicals).

Then all the input and output of the system has to be quantified. The detailed modeling under Aspen Plus, as previously presented in this chapter, is an interesting tool to give the complete and detailed life-cycle inventory (or mass and energy balances) on the whole bioenergy chain.

1.4.2. *Life-cycle assessment of heat production from biomass*

As in the previous study, the life cycle assessment of wood combustion for heat production for a house has been made [PEL 15]. The system boundaries were defined as: the forest growth, harvest, the transformation of wood into chips, logs and pellets, the transportation from the forest to the final user, and finally the combustion of wood in the different equipment. The manufacturing and transportation of the equipment itself were not included in the environmental analysis due to a lack of available data.

The main data for wood combustion technologies emissions are presented in Table 1.4. They are based on industrial experiments on wood combustion units [ROG 15].

Equipment	**Log stove**	**Pellet stove**	**Log boiler**	**Pellet boiler**	**Wood district heating**
Fumes exit temperature (°C)	330	125	150	160	110
Pollutant concentrations corrected for O_2 %vol. in smoke	13% O_2		10% O_2		11% O_2
Carbon monoxide (CO), mmol/Nm	361.61	22.32	133.93	3.18	8.625
Nitrogen oxides (NO_x), mmol/Nm	1.64	3.11	4.19	4.39	3.3
Particles, mg/Nm	42.04	48.16	7.66	99.44	10.0
Total volatile organic compounds (VOC), mmol/Nm	48.13	1.72	6.06	2.25×10^{-2}	9.07×10^{-2}
Methane (CH_4), Mmol/Nm	5.81	2.12×10^{-1}	1.81	8.13×10^{-3}	2.94×10^{-2}
Polycyclic aromatic hydrocarbons (PAHs), mmol/Nm	7.32×10^{-2}	3.65×10^{-3}	7.35×10^{-3}	3.47×10^{-5}	3.75×10^{-2}

Table 1.4. *Main emissions for wood combustion technologies used as input data for LCA [PEL 15]*

In an LCA framework, each input and output to the system is multiplied by an impact factor to obtain the total potential impact of the studied scenario. Although there is a general consensus about the impact of major pollutants such as CO_2, the impact factor for minor pollutants may vary according to the dataset. Here we only present one impact, the global warming potential (GWP). It was calculated for each scenario using three different methods of LCA (Table 1.5): ReCiPe Midpoint (Hierarchist) [PRÉ 15], EDIP2003 [POT 06] and CML2001 [DEB 02]. For electricity, the average French mix was used due to the high proportion of nuclear power in the French grid; the GWP impact of French electricity is somewhat lower than that of nearby Germany for example. The used impact factors are shown in Table 1.5. In order to compare the different scenarios, the functional unit is kWh of heat delivered to the user, that is either 1 kWh hot water for the heating district case or 1 kWh heat transferred from the stove to the room.

Flow	ReCiPe Midpoint (H)	EDIP2003	CML2001
CO_2	1 kg CO_2eq./kg	1 kg CO_2eq./kg	1 kg CO_2eq./kg
CH_4	25 kg CO_2eq./kg	23 kg CO_2eq./kg	25 kg CO_2eq./kg
CO		2 kg CO_2eq./kg	
N_2O	298 kg CO_2eq./kg	296 kg CO_2eq./kg	298 kg CO_2eq./kg
$C_{16}H_{10}$ (PAH surrogate)			16.1 kg CO_2eq./kg
Production and combustion of natural gas	0.198 kg CO_2eq./kWh	0.198 kg CO_2eq./kWh	0.198 kg CO_2eq./kWh
Electricity (French mix)	0.180 kg CO_2eq./kWh	0.180 kg CO_2eq./kWh	0.180 kg CO_2eq./kWh

Table 1.5. *Impact factors for different LCIA methods*

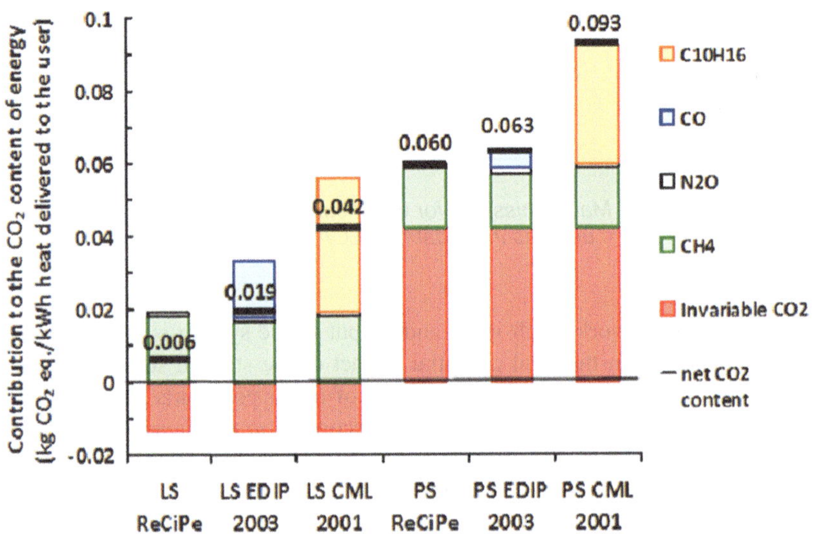

Figure 1.22. *Impact of minor pollutants on the GWP result, according to the three LCA methods (Table 1.5), for log (LS) and pellet stoves (PS) [PEL 15]. Each of the columns presents the different contributions to global warming expressed. The figure on top of the column is the total global warming estimated. For a color version of the figure, see www.iste.co.uk/dufour/biomass.zip*

The three methods do not account for the same impact (in terms of kg CO_2eq., see Table 1.5) for the various compounds emitted during wood combustion and this discrepancy can increase or decrease the calculated GWP impact by a factor 7 as highlighted in Figure 1.22.

Figure 1.22 highlights the huge uncertainty of the quantification of environmental impacts. The main issue to reduce this uncertainty is to specify the emissions of pollutants (by analysis of industrial units and Aspen Plus modeling) and to better quantify the impact factors for each pollutant species.

Figure 1.23 presents the cost of heat for the final user as a function of GWP impact of the different heating systems. It shows that all wood combustion systems result in a lower GWP impact than gas or electricity. Wood combustion associated with a district heating would be the cheapest option and pellet boiler the most expensive option.

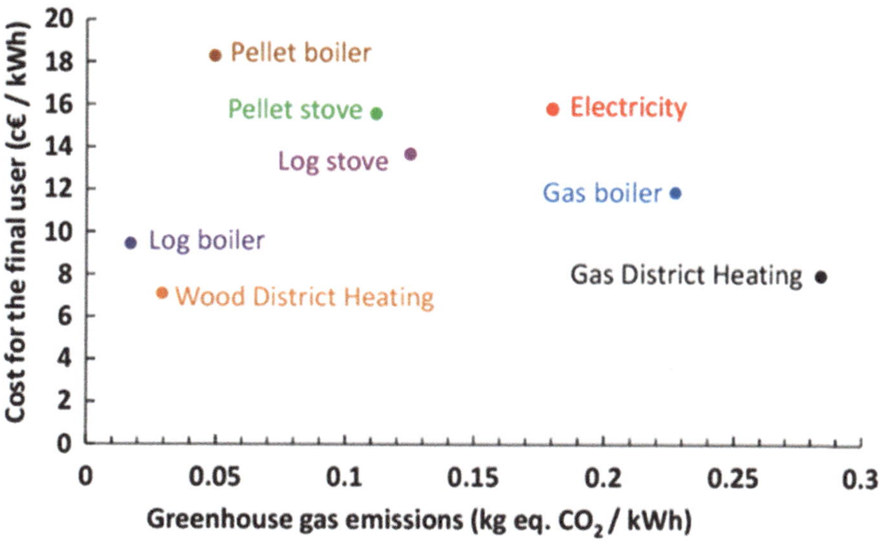

Figure 1.23. *Cost of heating technologies as a function of their greenhouse gas emissions [PEL 15]*

1.4.3. *Various aspects of the bioenergy chain*

Beyond the aspects of "physical" flows (mass and energy balances and LCA), various other aspects have to be considered in the bioenergy chain such as its impact on the society (employment, quality of life, health, etc.) (Figure 1.24).

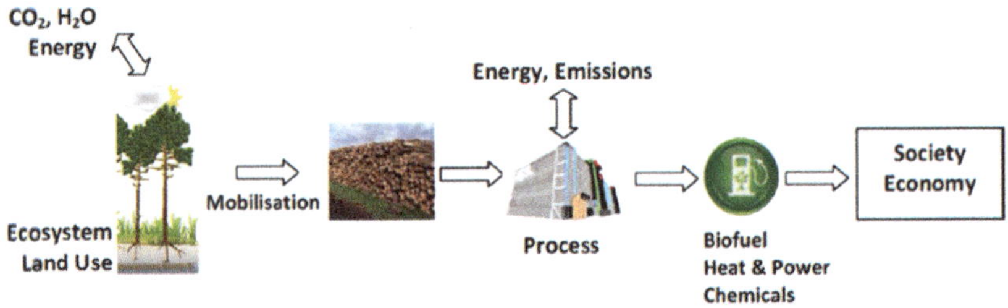

Figure 1.24. *Various aspects of the bioenergy chains: from the ecosystem to the society and global world economy*

Bioenergy can promote rural employment compared to fossil resources (Table 1.6).

Chain	Job created for 1,000 toe supplied
Forest biomass energy	4.2–6.3
Wood waste	2.3–3.7
Oil	1.4
Gas	1.2

Table 1.6. *Forest biomass energy creates more local employments than fossil resource*

For instance, the production of cleaner wood cookstoves for developing countries may promote the employment of women (in the cookstove and fuel value chain), saves time for the consumers (lower needs in energy wood), favor education, etc. It will also impact child development and survival due to a better indoor air quality [ANE 13].

1.4.4. *An example of a multicriteria discussion on a bioenergy route: the sequestration of biochar to mitigate climate change*

It has been shown that limiting anthropogenic climate change requires developing new routes for energy supply with net negative emissions, such as bioenergy with carbon capture and storage (BECCS) [VAN 13]. To achieve this goal, a large amount of work has dealt with the land application of biochar [LEH 06] or the sequestration of CO_2 from biomass combustion [Van 13]. In our opinion, the concept of wood geostorage [KRE 09, ZHE 08] has not yet received enough attention. We have proposed to extend this concept to the geological sequestration of biochar (GSB), displayed in Figure 1.25 [DUF 13a].

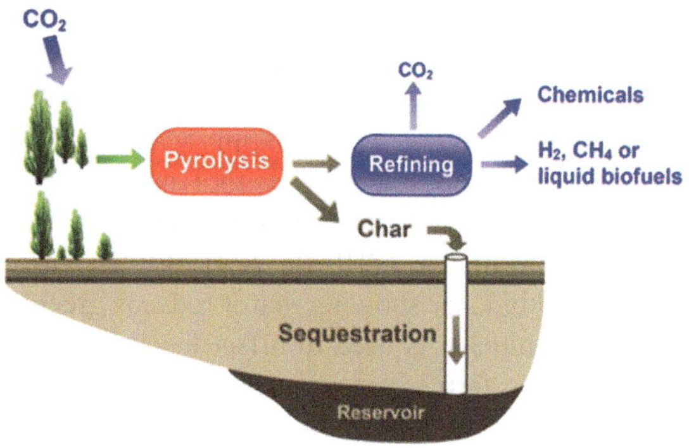

Figure 1.25. *Diagram of the concept of GSB. Biochar, a partly decarbonized energy vector (H_2, CH_4 or liquid biofuels), and added-value chemicals (e.g. aromatics) could be produced by a pyrolysis process. Biochar would be stored in geological cavities such as former coal mines*

1.4.4.1. *About biosphere management and the potential of carbon sequestration by biomass*

The potential of biomass for energy and carbon sequestration depends on critical uncertainties with respect to land and biomass availability, technology development and economical or societal aspects. Long-term primary forests have to be absolutely preserved for biodiversity, cultural heritage and landscape preservation. Other land could be used to manage the carbon pool of the biosphere, with fast-growing crops (such as *Miscanthus* and *Eucalyptus*) or small stems harvested to reduce stem density and to promote a sustainable management of forests (e.g. Douglas fir, pines). An advanced integrated model has shown that the total potential for negative emissions would be in the order of 10 $GtCO_2$/year for BECCS (with biogenic CO_2 sequestration) [VAN 13]. Kreysa [KRE 09] showed by assuming a reforestation scenario that the atmospheric CO_2 concentration could be lower than 400 ppm in the year 2100 (with a temporary maximum of 444 ppm around 2060) due to wood geostorage instead of 550ppm in 2100. Lehmann *et al.* [LEH 06] estimated a potential of 5.5–9.5 GtC/year for biochar sequestration by land application. These results differ depending on the model used and on the assumptions made, but they all highlight the big potential of carbon sequestration by biomass.

1.4.4.2. *GSB compared to wood geostorage and to land application of biochar*

GSB is a true carbon sink, the forest being the "pump and concentrator" of atmospheric CO_2 and the pyrolysis process the second "concentrator" of carbon into char. This concept is close to the geostorage of wood [ZHE 08, KRE 09] but different by two important features: (1) char is more stable and concentrated in carbon (mass of carbon per volume unit) than biomass and (2) a fraction of biomass is valorized as a (partly) decarbonized energy vector (such as CH_4, H_2 and methanol) and chemicals, leading to fossil fuel substitutions. These two features enhance the benefits argued for wood geostorage [ZHE 08, KRE 09]. Extensive work has dealt with the land application of biochar showing that it reduces greenhouse emissions to a greater extent than when biochar is used to offset fossil fuel emissions [LEH 06]. Biochar in soils also reduces fertilizer consumption and N_2O emission [LEH 06]. Nevertheless, the carbon in char is not stored in perpetuity when applied to land and it can affect the soil microbial activity. The land application is an interesting strategy though probably with a lower potential for long-term carbon sequestration than GSB. Furthermore, the char stored in geological reservoir such as former coal mines could be recovered whereas it seems hardly feasible after land application. In my opinion, these two sequestration methods could be combined and complementary.

1.4.4.3. *Potential societal and economic benefits of GSB*

The harvested biomass would be converted to wood chips and then transported to a medium-scale (<50 MW) pyrolysis unit. This medium scale should promote the local use of biomass, reduce transport costs and generate employment in rural areas or remote sites. The coproduction of syngas and added-value chemicals (e.g. BT) would offset a part of the lost revenue from the char not valorized. A small fraction of char could also be used for producing activated carbons to increase revenue. Syngas could be further converted to energy vectors or chemicals, for example H_2, CH_4 or a liquid biofuel. The capital cost and revenue will highly depend on site-specific parameters: the cost of biomass, electricity and fossil fuel, market need and price of chemical commodities, etc. The production of biobased chemicals, with a negative carbon footprint, would be a strong asset of the proposed process. The sequestered char would not be lost, but stored in "bio-mines." This carbon source remains available for future generations in case of energetic crisis or when new technologies will be available to mitigate the climate change. The societal acceptance is an important constraint for carbon sequestration. There may be a better perception by local communities for biochar sequestration, which is a recoverable stable solid, than for CO_2 sequestration from fossil fuel valorization.

1.4.4.4. *Technical and scientific perspectives*

The proposed concept would address some technical and scientific issues that have to be balanced with the relative complexity of the recovery, separation and sequestration of CO_2 from fossil fuel combustion! Batch or continuous processes are under operation for char production, for instance for the steel manufacturing industry. The selectivity in terms of gas, char and aromatic chemicals can be adjusted as a function of pressure, temperature and residence times in the reactors. Considering a pyrolysis at 700°C final temperature, the carbon in wood will be transferred for about 35% into char, 55% into permanent gas and 10% into aromatic compounds [DUF 09a]. Wood chips could be pyrolyzed under high-pressure conditions to reduce the conversion time, to promote exothermic reactions and thus to increase the energy balance and the char mass yield. The cracking of hydrocarbons over char could be used to fill the pores of biochar with a pyrolytic carbon and to promote H_2 production. Experiments with low-cost catalysts are needed to optimize the selectivity in added-value aromatics. Char could then be stored (as pellets, powder, etc.) in former coal mines or in other geological cavities. Long-term stability and geostorage studies have to be conducted, but a biochar would exhibit long-term stability in geological reservoirs.

2

Reactor Scale

2.1. Introduction about the reactions occurring in biomass thermochemical reactors

The study at the bioenergy chain and process scale points out that a better knowledge of reactions and reactors is needed, especially, for instance, to reduce the emissions of pollutants. The main reactions occurring in pyrolysis, gasification or combustion reactors are presented in Figure 2.1.

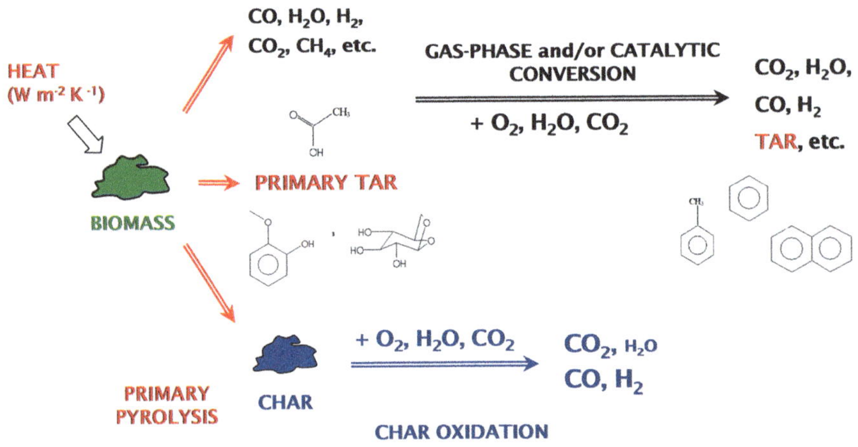

Figure 2.1. *Main reactions occurring in biomass pyrolysis, gasification and combustion reactors*

In all the thermochemical reactors, biomass is first heated by the reactor (by a flame or heat carrier material).

Upon heating, biomass is first dried and then decomposed by primary pyrolysis (150–500°C) into volatiles (permanent gas, water and primary tars) and char. Primary tars are composed of hundreds of molecular species produced by the thermal decomposition of biomass macromolecules. These molecules are composed of methoxyphenol derivatives (from lignin decomposition) and ketones, ethers, aldehydes, pyrans and furan derivatives (from cellulose and hemicelluloses decomposition). Then, the volatiles can undergo secondary reactions at the gas phase or over a catalyst. Primary oxygenated tars can be converted into secondary (phenolics and olefins) and tertiary tars (methyl aromatics and polycyclic aromatic hydrocarbons (PAH)) as a function of thermal severity (temperature × gas mean residence time) in the reactor [EVA 87].

The pyrolysis can be slow or fast depending on the heat flux density (W/m^2) brought by the reactor to the particles (Figure 2.2).

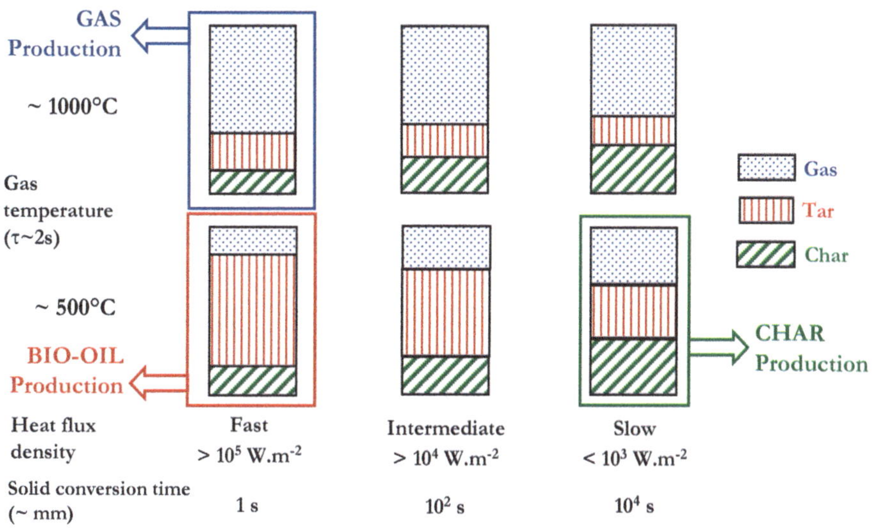

Figure 2.2. *The pyrolysis regime as a function of heat flux density brought by the reactor to the solid biomass and of temperature of the gas phase in the reactor (adapted from [DEG 82])*

Slow pyrolysis will promote char production. Fast pyrolysis (high heat flux density) at ~500°C gas-phase temperature will promote production of bio-oils. Fast pyrolysis at high gas-phase temperatures will favor the formation of gas from tar cracking. Heat flux density depends on the type of reactor and is equal to the external heat transfer coefficient (W/m^2K) multiplied by the difference in temperature between the heat source in the reactor and the surface of the particle (for instance, 480 K for a particle at 20°C in a reactor at 500°C). External heat transfer for fluidized bed and convective hot gas are in the range of 500 and 50 W/m^2 K, respectively (Figure 2.3).

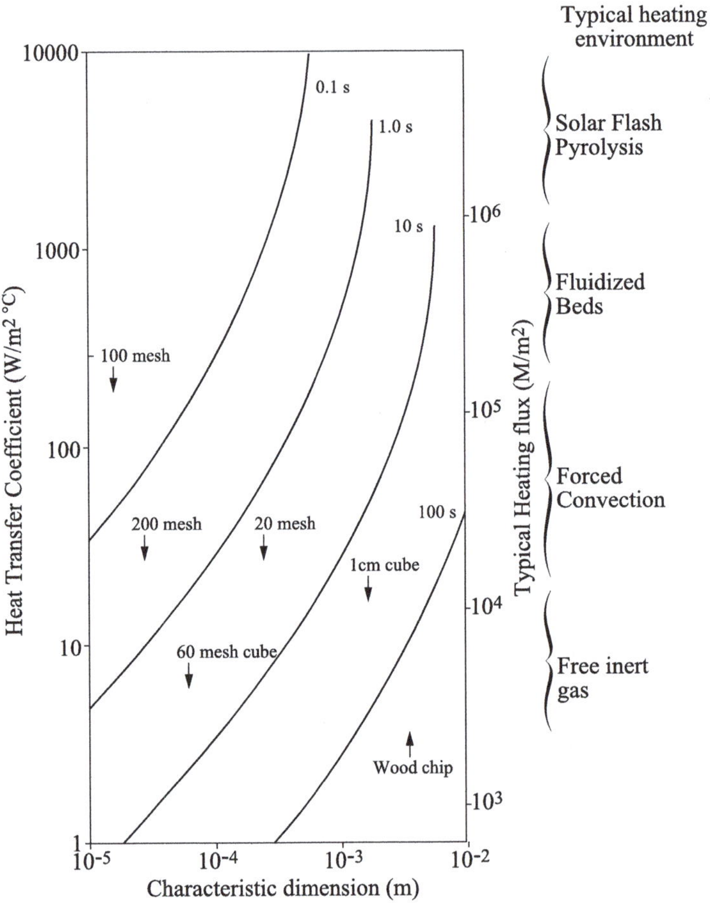

Figure 2.3. *Heat transfer coefficient of typical heating environment and heating time as a function of particle characteristic dimension [GRA 84]*

In the fast pyrolysis reactors, the vapor residence time should be reduced to its minimum (<2 s at 500°C) to reduce secondary reactions and to promote the formation of primary tars, which lead to the production of "bio-oils" after condensation.

In the gasification reactors, tars have to be converted into syngas by the addition of a catalyst or by high-temperature gas-phase conversion (>800°C). Tertiary tars are undesired products in the gasification systems.

In the combustion reactors, the very high temperature in flame obtained by the combustion of pyrolysis gas with air results in an almost complete oxidation of pyrolysis products (gas and tar) into CO_2 and H_2O. Nevertheless, there are still pollutants formed by these oxidation reactions (mainly aromatic compounds such as benzene and PAH).

An oxidant gas (O_2, H_2O and CO_2) is fed to the reactor to promote tar conversion and also to oxidize char. The oxidation by O_2 of pyrolysis products will bring the heat required for the reactors.

All these reactions can occur in the same reactor or can be decoupled in staged reactors for improving the selectivity in a given goal product (e.g. syngas) [ZHA 13].

In this chapter, we show how these reactions are combined in the reactors.

2.2. Reactors for biomass combustion

2.2.1. *Basics of biomass composition, drying and combustion*

The main combustion parameter is the excess air ratio (λ) that describes the ratio between the locally available and the stoichiometric amount of combustion air.

For biomass, the global combustion reaction can be described [NUS 03] as follows:

$$\begin{aligned} & CH_{1.44}O_{0.66} + \lambda 1.03(O_2 + 3.76\,N_2) \\ & \quad \rightarrow \text{intermediates } (C, CO, H_2, CO_2, C_mH_n, \text{etc.}) \\ & \quad \rightarrow CO_2 + 0.72\,H_2O + (\lambda - 1)O_2 + \lambda 3.87\,N_2 \qquad 439\ \text{kJ/kmol} \end{aligned}$$

if heteroatoms present in biomass (N, K, Cl, etc.) are neglected. Here $CH_{1.44}O_{0.66}$ represents the average composition of typical biomass (wood, straw, etc.); the intermediates represent the pyrolysis products with hundreds of molecules.

The pollutants are also not described by this simplified equation. CO, PAH, soot, N and S species, etc. can be formed during pyrolysis and/or the incomplete oxidation of pyrolysis products.

Ashes are also formed during combustion from the conversion of nutrients (species composed of Si, K, Ca, etc.) present in wood.

The main issue for the combustion reactors is to control the reactions for reducing the formation of these pollutants.

The higher heating value (HHV) of wood corresponds to the heat produced by the complete oxidation of one mass unit of fuel, with water (present in flue gas) condensation. It can be calculated by the following equation:

$$\mathrm{HHV} = 12.147xc - 0.097$$

where HHV is in kWh/kg and xc is the mass fraction of carbon in biomass.

The compositions of some biomasses are given in Table 2.1.

% mas.	Pine	Eucalypt	Oak	Willow	Wood average composition	Wheat straw
C	51.1	52.8	51.1	48.7	**51.0**	49.5
H	6.0	6.4	5.3	5.9	**6.0**	5.6
O	42.8	40.4	42.7	44.5	**42.5**	43.3
N	0.1	0.4	0.9	0.9	**0.5**	1.1
S	-	-	-	0.045	-	0.1
Cl	-	-	-	0.016	-	0.35
Ashes	0.4	0.8	2.7	2.1	**1**	7–16

Table 2.1. *Mass fractions in main constituents of biomasses*

All biomasses have similar carbon mass fraction (on an anhydrous basis). Therefore, they all have similar HHV between 4.5 and 6 kWh/kg. The main difference between biomasses in terms of calorific value is their density; thus, the HHV by volume unit (for instance, for one log of wood) is different.

The lower heating value (LHV) is more often used and corresponds to the heat (J or kWh) produced by the complete oxidation of a fuel, without water in the flue gas condensation. It can be calculated by the following formula:

$$\mathrm{LHV}_{\text{anhydrous biomass}} = 9.45xc + 33.79xh - 3.48xo$$

where LHV is in kWh/kg, xh is the mass fraction of hydrogen and xo is the mass fraction of oxygen.

The LHV of wet biomass depends on its moisture content:

$$H_D = 100\frac{m_{wet} - m_{anhydrous}}{m_{anhydrous}} \qquad H_W = 100\frac{m_{wet} - m_{anhydrous}}{m_{wet}}$$

where H_D is the biomass moisture content on dry biomass basis; and H_W is the biomass moisture content on wet biomass basis.

Wet biomass LHV reads:

$$LHV_{wet} = (1 - H_W/100)LHV_{anhydrous} - (H_W/100)Lv$$

where Lv is the heat of water vaporization (0.625 kWh/kg at atmospheric pressure).

Figure 2.4 presents the effect of moisture content in wood on its LHV. This highlights the importance of biomass drying for combustion, gasification and pyrolysis process (for "dry" thermochemical processes). For hydrothermal liquefaction (conversion of biomass in a liquid water), the high water content in biomass is not an issue.

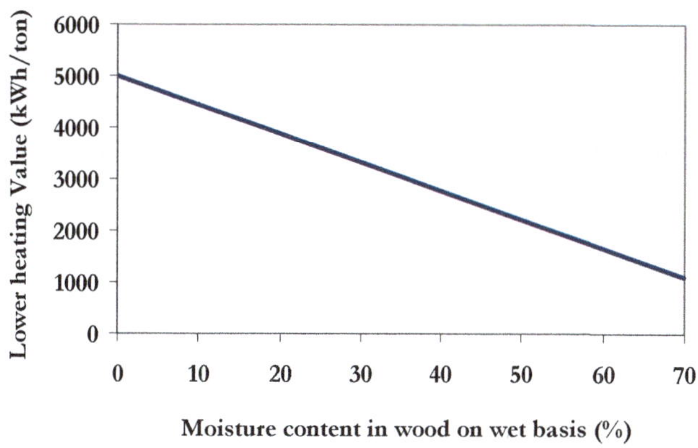

Figure 2.4. *Evolution of LHV as a function of moisture content in wood*

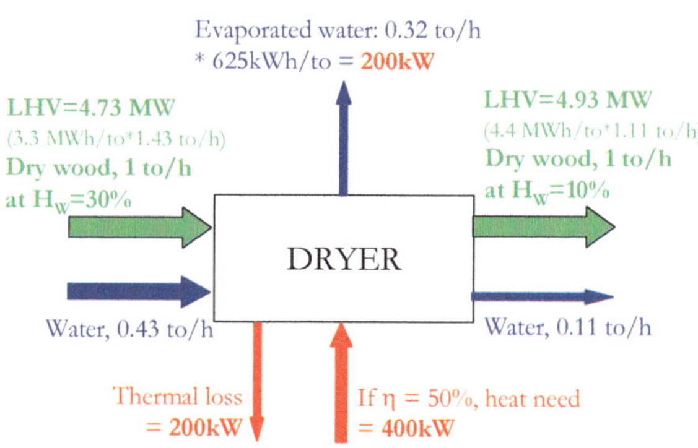

$$M_{H2O} = M_{drywood} * \left(\frac{1}{1-\frac{H_W}{100}} - 1\right)$$

Figure 2.5. *Simplified mass and energy balances of a dryer. For a color version of the figure, see www.iste.co.uk/dufour/biomass.zip*

Figure 2.5 gives simplified mass and energy balances for wood drying. Drying is not the main scope of this book and will not be described further.

2.2.2. *Technologies of biomass combustion*

Various cookstove technologies have been developed, especially in developing countries (see Figure I.3 in Introduction of this book) [KSH 14]. The technologies of biomass combustion have been comprehensively reviewed by Nussbaumer [NUS 03].

The main concept to improve combustion is the staging of air or fuel (Figure 2.6) being in small cookstove or in industrial boilers.

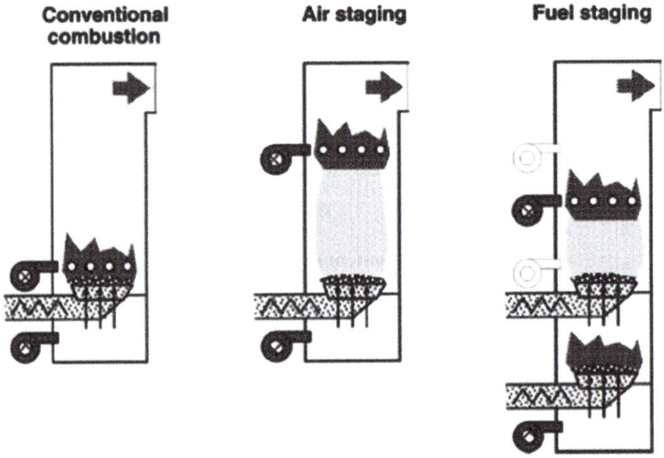

Figure 2.6. *Principle of conventional staging (air or fuel) during combustion [NUS 03]*

Air staging enables good mixing of combustion air with the combustible gases formed by pyrolysis and gasification of the fuel bed. This good mixing enables an operation at low excess air (λ < 1.5) and, thus, a higher efficiency and high temperature. Therefore, the concentrations of unburnt pollutants can be reduced. Staging has been also developed to reduce the formation of fuel NO_x (NO_x formed from the oxidation of N species present in biomass). Figures 2.7 and 2.8 present some combustion technologies. Downdraft boilers (Figure 2.7) with inverse flow have been developed to achieve a high efficiency in small-scale combustion based on the two-stage combustion principle.

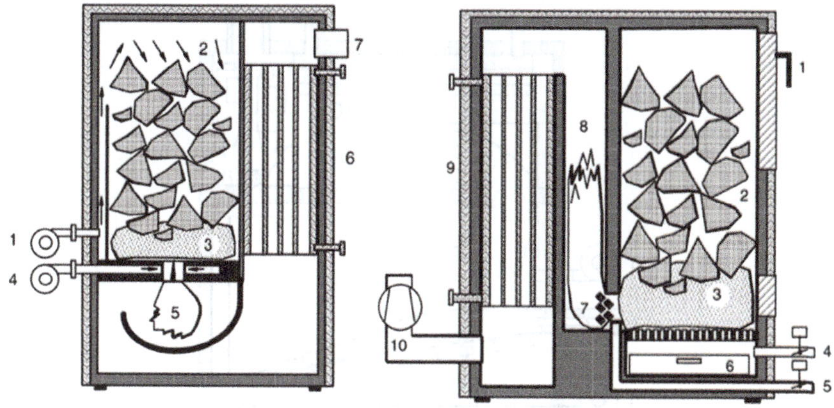

Figure 2.7. *Downdraft boilers with inverse combustion of wood log and with enforced air flow and air supply with primary and secondary air. Left: 1, primary air; 2, fuel hopper; 3, glow bed; 4, secondary air; 5, postcombustion; 6, heat exchanger; 7, chimney. Right: 1, fuel inlet; 2, fuel hopper; 3, glow bed; 4, primary air; 5, secondary air; 6, ash bin; 7, mixing zone; 8, postcombustion chamber; 9, heat exchanger; 10 chimney [NUS 03]*

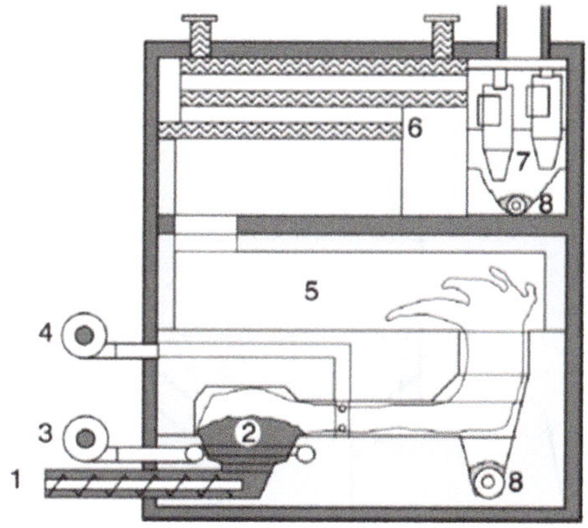

Figure 2.8. *Understoker furnace with primary and secondary air, mixing zone and postcombustion chamber: 1, screw feeder; 2, understoker zone with glow bed; 3, primary air; 4, secondary air; 5, postcombustion chamber; 6, heat exchanger; 7, cyclone; 8, ash removal [NUS 03]*

Wood pellets are relatively homogeneous fuel with a low water content; therefore, they are well suited for automatic heating at small scales.

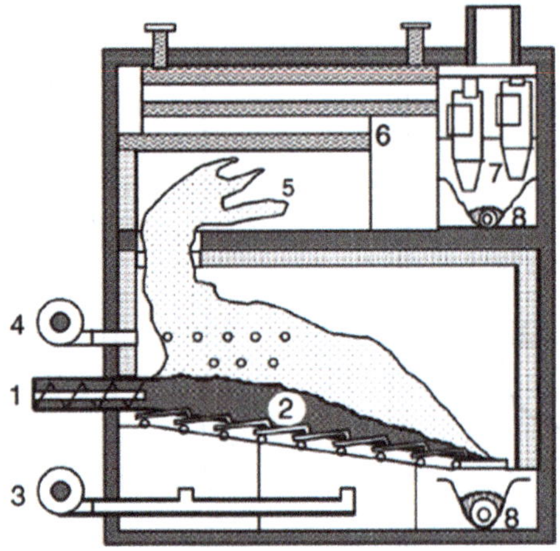

Figure 2.9. *Moving grate furnace with primary air in two stages in the grate and secondary air. 1, screw feeder; 2, moving grate; 3, primary air; 4, secondary air; 5, postcombustion chamber; 6, heat exchanger; 7, cyclone; 8, ash removal [NUS 03]*

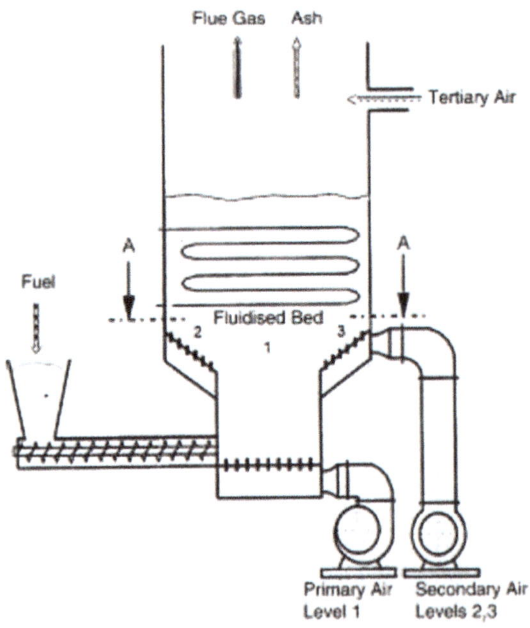

Figure 2.10. *Dense (or bubbling) fluidized-bed combustor with stage combustion (Lurgi) [NUS 03]*

Understoker furnaces are mostly used for wood chips with a low ash content (few agglomeration troubles) (Figure 2.8). Grate furnaces are better applied for high ash and water content biomasses (such as bark) (Figure 2.9). Dense fluidized beds (Figure 2.10) (with a low fluidization air velocity) or circulating fluidized bed (with a high fluidization air velocity) are used for large-scale applications, for instance for wood waste or waste from the pulp and paper industry. In fluidized-bed reactors, the high mixing of sand enables a nearly homogeneous temperature and, thus, a high burnout quality at a low excess air [NUS 03].

Table 2.2 summarizes the various combustion technologies for each type of wood fuels.

application	type	typical size range	fuels	ash	water content
manual	wood stoves	2 kW–10 kW	dry wood logs	<2%	5%–20%
	log wood boilers	5 kW–50 kW	log wood, sticky wood residues	<2%	5%–30%
pellets	pellet stoves and boilers	2 kW–25 kW	wood pellets	<2%	8%–10%
automatic	understoker furnaces	20 kW–2.5 MW	wood chips, wood residues	<2%	5%–50%
	moving grate furnaces	150 kW–15 MW	all wood fuels and most biomass	<50%	5%–60%
	pre-oven with grate	20 kW–1.5 MW	dry wood (residues)	<5%	5%–35%
	understoker with rotating grate	2 MW–5 MW	wood chips, high water content	<50%	40%–65%
	cigar burner	3 MW–5 MW	straw bales	<5%	20%
	whole bale furnaces	3 MW–5 MW	whole bales	<5%	20%
	straw furnaces	100 kW–5 MW	straw bales with bale cutter	<5%	20%
	stationary fluidized bed	5 MW–15 MW	various biomass, $d < 10$ mm	<50%	5%–60%
	circulating fluidized bed	15 MW–100 MW	various biomass, $d < 10$ mm	<50%	5%–60%
	dust combustor, entrained flow	5 MW–10 MW	various biomass, $d < 5$ mm	<5%	<20%
co-firing[a]	stationary fluidized bed	total 50 MW–150 MW	various biomass, $d < 10$ mm	<50%	5%–60%
	circulating fluidized bed	total 100–300 MW	various biomass, $d < 10$ mm	<50%	5%–60%
	cigar burner	straw 5 MW–20 MW	straw bales	<5%	20%
	dust combustor in coal boilers	total 100 MW–1 GW	various biomass, $d < 2–5$ mm	<5%	<20%

[a] Biomass covers typically less than 10% of the total fuel input.

Table 2.2. *Types of biomass furnaces with typical applications and fuels [NUS 03]*

2.3. Reactors for biomass gasification

2.3.1. *Main applications for syngas produced from biomass gasification*

Figure 2.11 presents the main applications for the valorization of gas produced from the gasifiers. The main issue for all these processes is the purification of syngas and mainly the removal of tar to meet the specifications of the gas engine or catalytic reactors for methanation or Fischer–Tropsch (FT) synthesis. The detailed specifications for syngas purity are not yet well known due to the lack of industrial data for catalytic synthesis (long-term experiments at industrial scale). The efficiencies of these processes are also given in Figure 2.11.

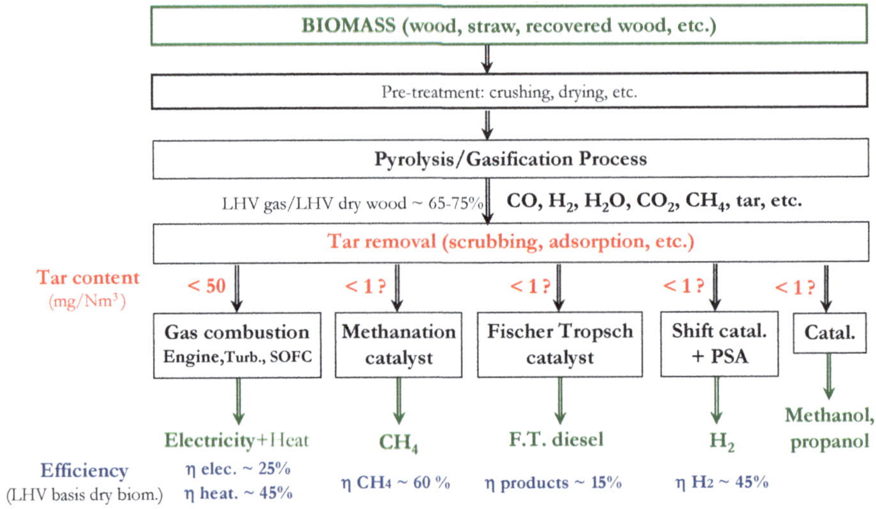

Figure 2.11. *Main routes for syngas valorization*

2.3.2. *Different type of reactors for biomass gasification*

The first type of gasification reactors are fixed-bed reactors (Figure 2.12). This technology has been developed since 1920 for wood (Imbert gasifier). In these reactors, the drying, pyrolysis, oxidation of pyrolysis products and gasification of char are stratified. The residence time of the solid is long (hundreds of seconds), so the reactions of char gasification ($C + H_2O = CO + H_2$, $C + CO_2 = 2CO$) are made possible (their kinetics are relatively slow at temperatures lower than 900°C).

In an updraft gasifier, biomass is injected from the top and air from the bottom. The producer gas is thus rich in tar (from pyrolysis zone). In a downdraft fixed bed, air is injected after the pyrolysis zone, and, therefore, the tar produced by pyrolysis is oxidized at high temperature and well cracked in the bed of char. The produced gas exits from the bottom of the bed and is cleaner (low tar content) than that in the updraft configuration. Main technology suppliers, drawbacks and advantages of these technologies are summarized in Figure 2.12.

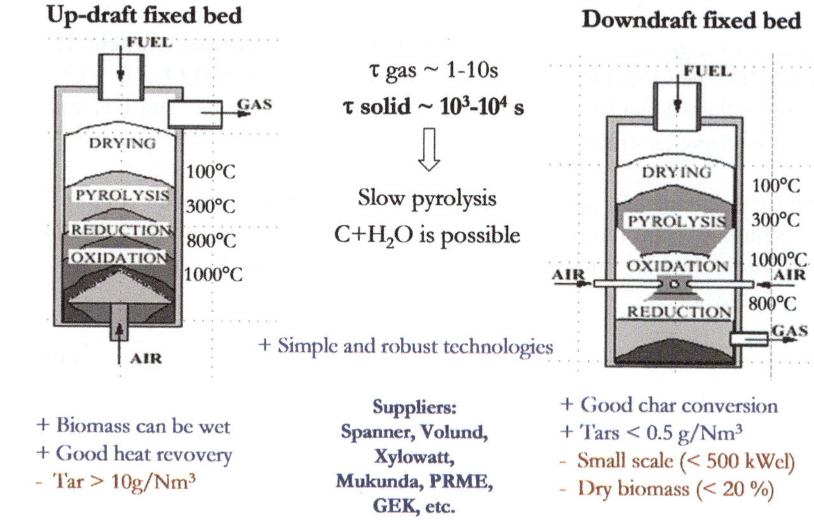

Figure 2.12. *Fixed-bed technologies for biomass gasification*

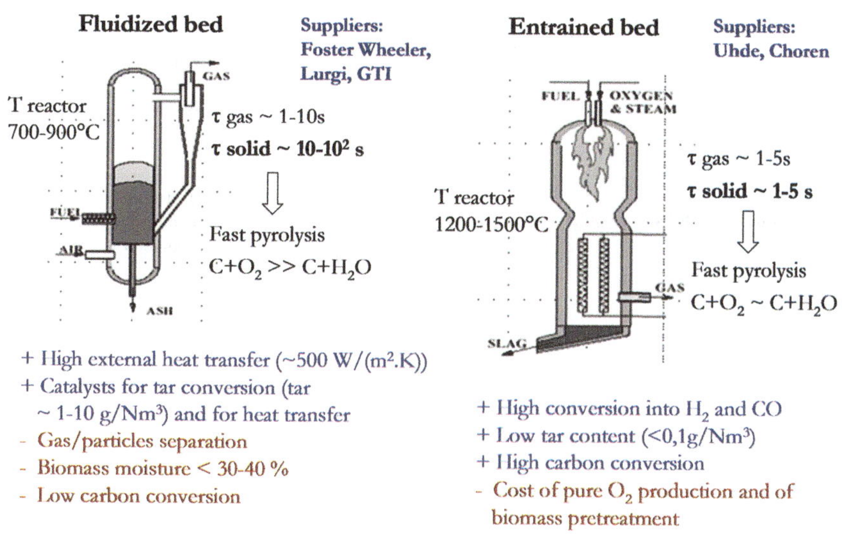

Figure 2.13. *Fluidized bed and entrained bed for biomass gasification*

The other technology is fluidized-bed reactors (Figure 2.13) for larger-scale units rather than fixed-bed reactors. As for combustion, the fluidized bed can be dense (or bubbling, low fluidization velocity) or circulated. In these reactors, pyrolysis is fast because of a good heat transfer from the hot sand to biomass particle. If air is used to fluidize the sand, the gasification reactions of char ($C + H_2O$ or $C + CO_2$) are much slower than those for char oxidation ($C + O_2 = CO/CO_2$). The bed can be a catalyst to improve tar cracking.

Another technology for very big scales is the entrained flow reactor (Figure 2.13). In this technology, the temperature of the reactor is very high, thus promoting tar cracking. But ashes are melted producing slags. This may be an issue for ashes recycling to the soil.

The main drawback of the air fluidized-bed gasifier is that the syngas is diluted in N_2 from air. To overcome this problem, dual fluidized-bed (DFB) gasifiers have been proposed (Figure 2.14). In a DFB, the pyrolysis of biomass and secondary reactions are conducted in a first fluidized bed (gasifier). Char is then injected into a second fluidized bed with sand for its oxidation by air (combustor). A DFB used for gasifying biomass with steam or recycled gas is considered as a promising technology in order to produce a high-quality gas (mainly CO and H_2) without dilution in nitrogen. The concept actually emerged in the 1970s from pioneering studies conducted by Battelle, Columbus, in the United States [FEL 88].

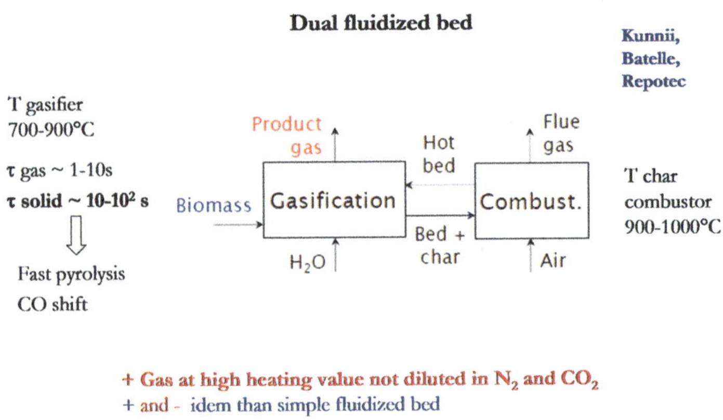

Figure 2.14. *Concept of dual fluidized-bed gasification. For a color version of the figure, see www.iste.co.uk/dufour/biomass.zip*

Depending on the technology of the DFB, the gasifier and the combustor can be a dense fluidized bed or a pneumatic riser (high velocity as in a circulating fluidized bed). The DFB technology can also depend on the level of the hot sand injection from the combustor to the gasifier. Indeed, the hot sand could be injected to the bottom, middle or top of the fluidized-bed gasifier.

Table 2.3 displays three technologies of DFB [ABD 12].

Name of the technology	**TNEE [DEG 91]**	**FERCO [FEL 88]**	**Güssing [PFE 04]**
Development beginning	In the 1980s	In the 1970s	In the 1990s
Simplified scheme	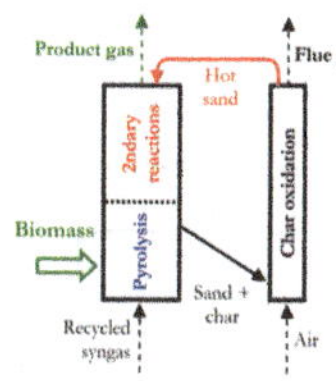	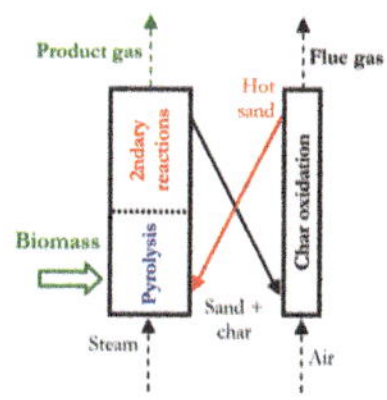	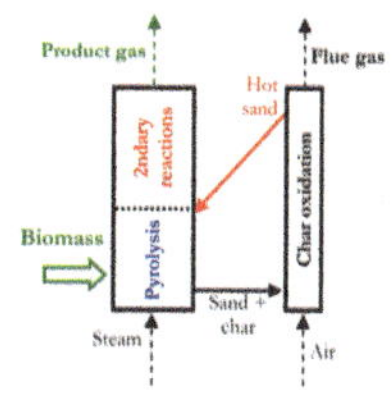
Description	Gasification: LVFB Char oxidation: HVPR The hot bed material is injected to the top of the pyrolyzer. Better gas/solid contact between the pyrolysis gas and the hot sand from combustion due to the "Saturne" contactor	Gasification: HVPR Char oxidation: HVPR Owing to HVPR gasification, the solid residence time is low and incomplete pyrolysis conversion could be achieved	Gasification: LVFB Char oxidation: HVPR The bed material is recycled to the middle of the gasifier. Little hot catalyst is present in the freeboard of the gasifier
Technology definition (see text for the definition of Y_i)	$Y_1 = 0.05$, $Y_2 = 0$, $Y_3 = 1$, $Y_4 = 1$	$Y_1 = 0$, $Y_2 = 1$, $Y_3 = 0$, $Y_4 = 0$	$Y_1 = 0.05$, $Y_2 = 0$, $Y_3 = 0$, $Y_4 = 0$

LVFB, low-velocity fluidized bed; HVPR, high-velocity pneumatic riser

Table 2.3. *Simplified diagram and description of three complementary technologies of DFB gasifier (biomass introduced by screw conveyors for all technologies.)*

The gas composition for each technology is presented in Table 2.4. The main differences between the gasification technologies lie in the tar and particle contents in the producer gas and on the dilution by N_2 of air. Dilution by N_2 directly impacts the heating value of the gas (LHV in Table 2.4).

	Unit	Downdraft Fixed bed	Updraft Fixed bed	fluidized bed	Dual fluidized bed (FICFB)	Entrained Bed (Future Energy)
Water content in biomass	% dry biom.	6-25	10-50	13-20	n.d. (~ 20)	Pretreated biomass
Particles	g/Nm^3	0.1-8	0.1-3	8-100	5-10	~ 0 (vitrification)
Tar	g/Nm^3	0.01-6	10-150	2-30	1.5-4.5	n.d. (~ 0)
LHV	kWh/Nm^3	1.1-1.6	1-1.9	1-1.6	3.6	2.6
H_2	% vol.	15-21	10-18	15-22	35-45	27
CO	% vol.	10-22	15-25	13-15	20-30	50
CO_2	% vol.	11-13	8-10	13-15	15-25	14
CH_4	% vol.	1-5	2-5.5	2-4	8-12	< 0.1
C_nH_m	% vol.	0.5-2	n.d.	0.1-1.2	n.d.	n.d. (~ 0)
N_2	% vol.	difference ~ 50	difference ~ 50	difference ~ 50	3-5	6

Table 2.4. *Composition of producer gas for the different technologies of gasification*

The selection of a gasifier technology depends on the scale of the unit and on the type of application for the gas. Table 2.5 presents the main technologies as a function of application and scale.

Reactor type	Application	Scale (ton dry biomass/h)
Downdraft	Power	<0.2
Fluidized bed	Power, CH_4, H_2, FT diesel	~1 to 50
Entrained bed	FT diesel, H_2	>10 (?)

Table 2.5. *Selection of gasifier technology*

DFB reactors are especially required for medium-scale applications (~50 MWth), especially for CH_4 production (substitute natural gas).

Large-scale units result in long transport distances of biomass. For this reason, a pretreatment step of biomass could be developed to improve the energy density of bioenergy.

Various pretreatment chains have been evaluated by Van der Drift *et al.* [VAN 04] (Figure 2.15).

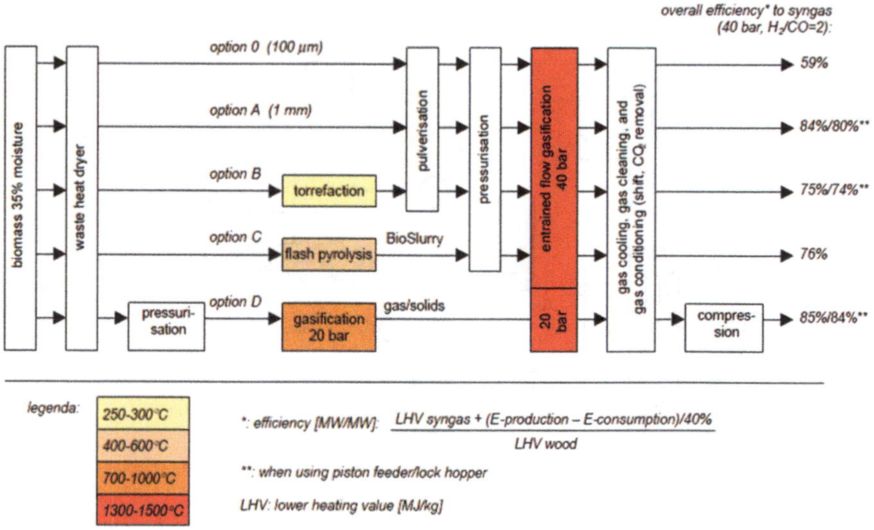

Figure 2.15. *Pretreatment of wood for gasification in large-scale entrained flow gasifier [VAN 04]. For a color version of the figure, see www.iste.co.uk/dufour/biomass.zip*

2.3.3. *Purification and valorization of syngas*

The main issue for all applications is the purification of syngas. Composition required for various applications are given in Table 2.6. There are still large uncertainties for these values because of the lack of long-term operations for catalytic synthesis at industrial scales.

Application	Tar (mg/Nm^3)	Particles (mg/Nm^3)	H_2S + COS	HCN–NH_3	HCl + HF
Gas engine	<50	<50	<80 mg/Nm^3	<50 mg/Nm^3	<100 mg/Nm^3
Methanation (Ni-based)	<5	<10	<100 ppb	<1 ppm	<25 ppb
FT synthesis (Fe/Co)	<0.1	<0.1	<10 ppb	<20 ppb	<10 ppb

Table 2.6. *Composition required for main syngas uses*

An example of a complete gasification unit (Güssing, Austria) is depicted in Figure 2.16. This process has been operated for several thousands of hours. It combines a DFB gasifier with a scrubber (for tar removal) and a gas engine.

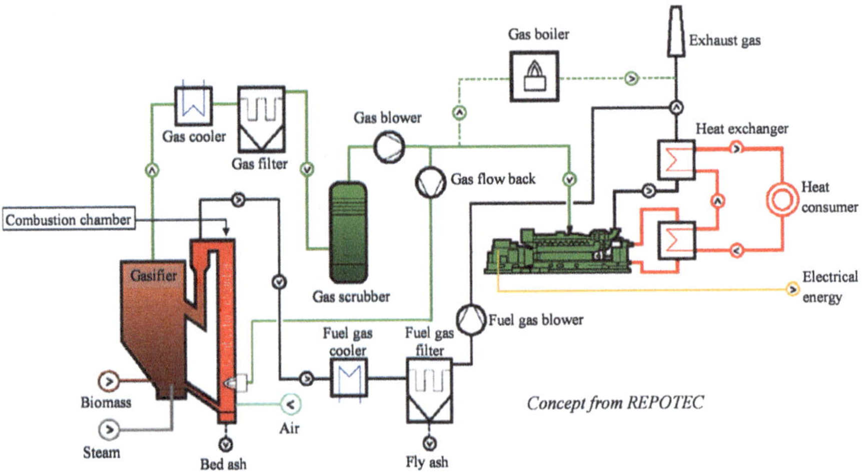

Figure 2.16. *The Güssing process for combined heat and power production [PFE 04]. For a color version of the figure, see www.iste.co.uk/dufour/biomass.zip*

Figure 2.17. *Tar deposit in syngas engines (Jenbacher)*

The problems with tar deposit in a gas engine (Jenbacher) if tar removal by the purification system is not sufficient are well depicted by Figure 2.17. Indeed, the tar dew point should be lower than 35°C to avoid any deposit in gas engine (Jenbacher).

Examples of gas cleaning trains for advanced (FT or methanation synthesis) are presented in Figure 2.18.

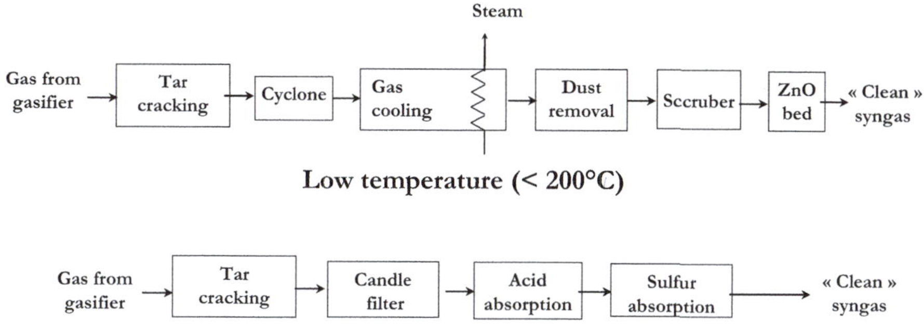

Figure 2.18. *Examples of gas cleaning trains (not comprehensive) for advanced gas applications [HAM 04]*

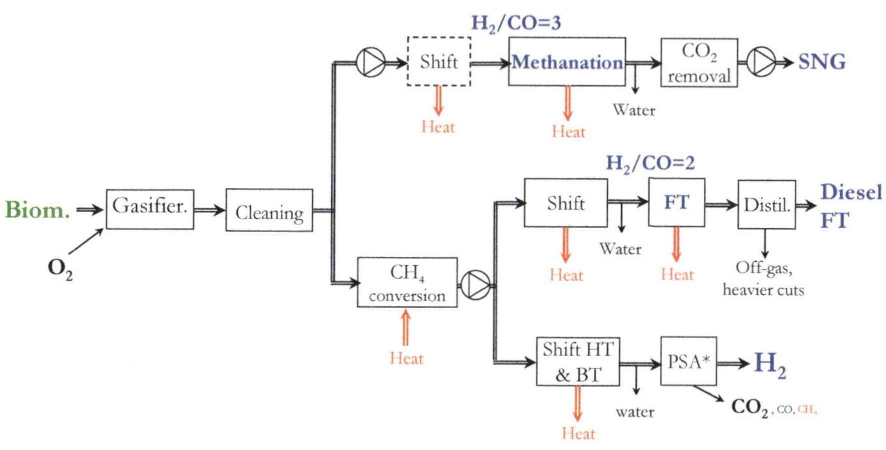

Figure 2.19. *Main steps for the synthesis of CH_4 (substitute natural gas (SNG)), diesel FT, or pure H_2. PSA, pressure swing adsorption*

The main steps (after syngas cleaning) for methanation, FT diesel or alcohols production are presented in Figure 2.19.

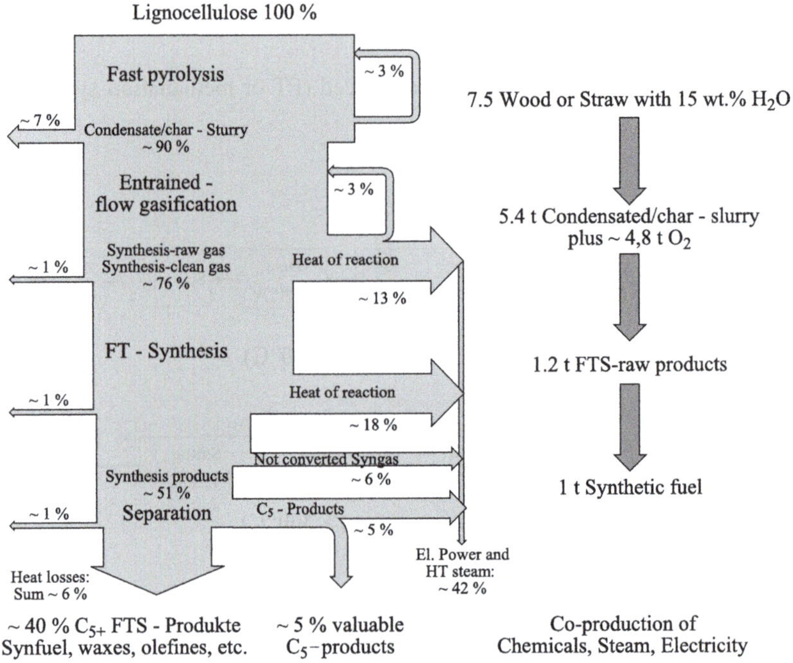

Figure 2.20. *Mass balance of gasification followed by Fischer–Tropsch synthesis [HEN 09]*

A mass balance of a process chain combining biomass pretreatment by fast pyrolysis, gasification of the bio-oil and char in an entrained flow gasifier and FT synthesis is presented in Figure 2.20.

2.3.4. *Detailed modeling of a dual fluidized-bed gasifier*

Our group has developed a detailed model of biomass gasification in a DFB reactor [ABD 12], also including the whole syngas purification and valorization [FRA 13a].

In our approach, the DFB is divided under Aspen Plus® into three modules according to the main chemical phenomena: biomass pyrolysis, secondary reactions and char combustion (Figure 2.21).

The different designs of DFB presented in Table 1.1 can be modeled by this decoupling approach and by modifying the mass ratio (Y_i in Table 2.3 and Figure 2.21).

The model of the TNEE technology has been used in the process and bioenergy chain models presented in Chapter 1.

Mass yields of permanent gases, water, 10 tar species and char are modeled with respect to the reactor temperature by a pyrolysis correlation obtained from our experimental work [DUF 09a]. The secondary reactions are modeled by a semi-detailed kinetic mechanism that handles gas-phase and catalytic conversions over char of CH_4 and lumped tar species (phenol, naphthalene, benzene and toluene), gas-phase water–gas shift reaction (WGSR) ($CO + H_2O = CO_2 + H_2$), char and soot steam gasification. The kinetic laws of about 15 reactions were selected based on the literature and on our experimental results. The hydrodynamics in the DFB reactor is modeled in a very simplified way (see Figure 2.21).

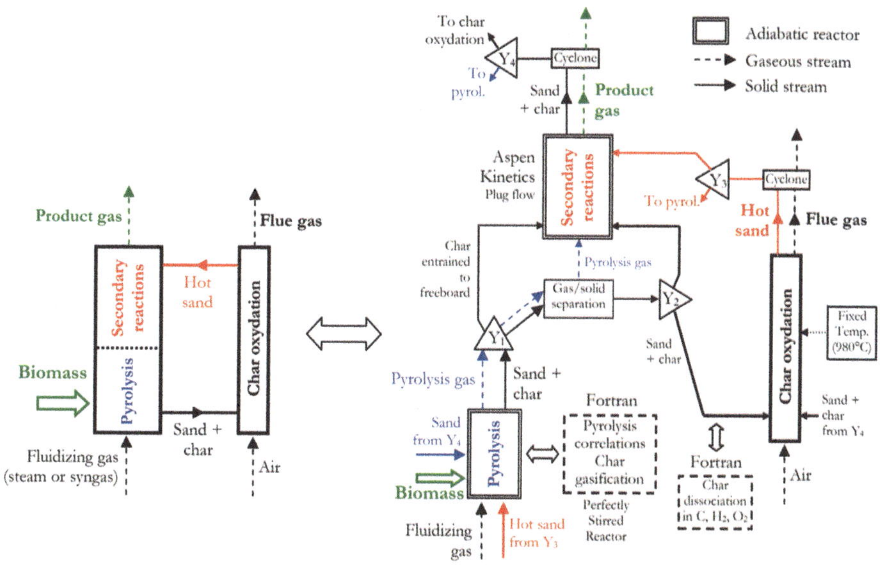

Figure 2.21. *Approach for modeling DFB gasifier under Aspen Plus software [ABD 12]*

The calculated compositions of permanent gases and tars, flow rates and LHVs of gas were compared with experimental data for two DFB technologies (TNEE and FERCO). Figures 2.22 and 2.23 show the comparison between experimental and predicted gas and tar compositions.

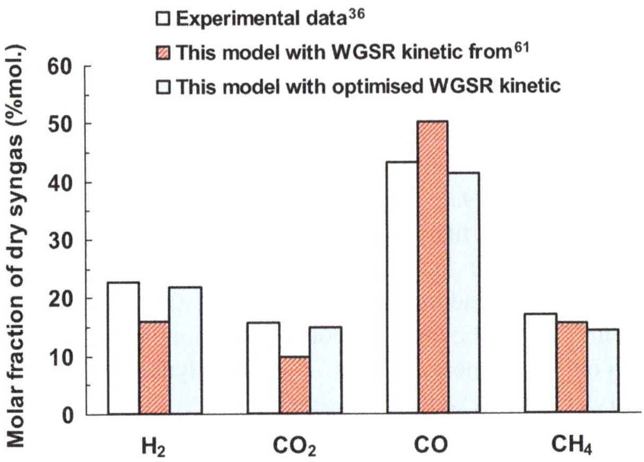

Figure 2.22. *Comparison between model predictions and experimental data obtained on TNEE pilot plant for gas composition*

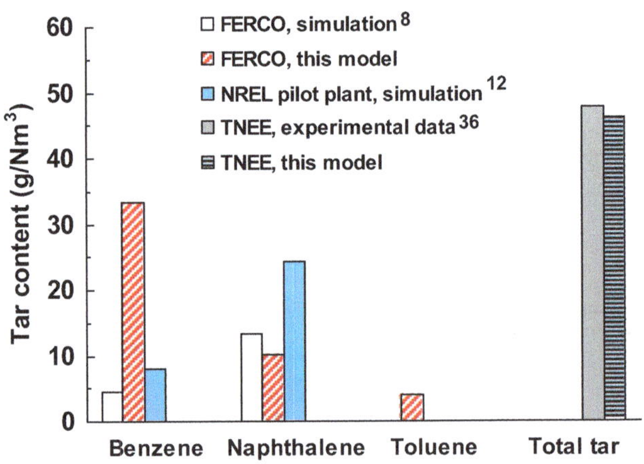

Figure 2.23. *Comparison between model predictions and experimental data obtained on TNEE and FERCO pilot plants for tar content in the syngas*

The syngas composition and flow rate are very sensitive to the WGSR kinetics. The rate laws for WGSR were reviewed, and an optimized kinetic law for WGSR was given.

Our model prediction always gives a higher benzene content compared with that provided by NREL correlations [SPA 05]. Only a slight difference appears for naphthalene content between Spath *et al.* [SPA 05] correlation and our work.

Spath *et al.* assumed in their model that 25 and 75 wt.% of tar content are benzene and naphthalene, respectively, which seems less realistic than our pyrolysis correlations. Benzene is the major tertiary tar component [CAR 07]. Tars were analyzed in the FERCO experiments by gravimetric method after separation from the quench water and followed by distillation [FEL 88]. Tar content was surprisingly independent of the reactor temperature between 0.9 and 1.5 wt.% on dry biomass basis, most probably because of artifacts from the tar analysis method. A new NREL correlation [KIN 09] was obtained from the experiments on the NREL pilot plant [JAB 09]. These new correlations have been determined with a more advanced tar analysis method in an extensive work but on another fluidized-bed technology [JAB 09]. The naphthalene content is underpredicted by our model compared with the new correlation [KIN 09]. This discrepancy could not be explained yet. According to our measurement [DUF 09], it seems more realistic to get a higher yield of benzene than of naphthalene. Benzene is much more produced by pyrolysis and secondary reactions and is more stable than naphthalene [JES 96].

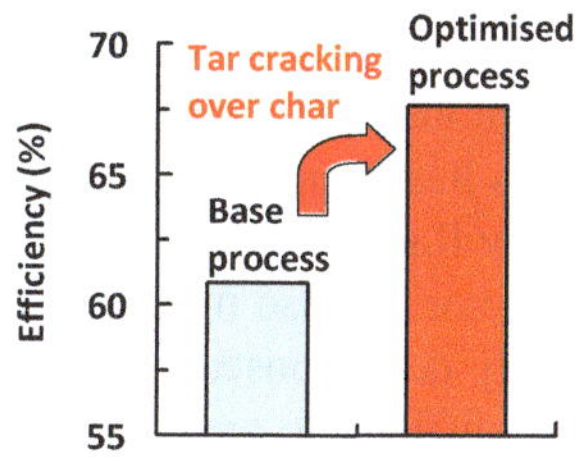

Figure 2.24. *Increase in cold gas efficiency predicted by the model when the contact time of tar over hot char (~900°C) is increased in the secondary reactions zone*

This detailed model that includes catalytic conversion of tar over the hot char has been used to model the effect of an increased gas/hot char contact time in the secondary reactions zone, for instance by an increase in the volume of the secondary reaction zone and/or char concentration in the hot sand. If tar conversion over hot char is promoted, the cold gas efficiency (LHV of gas/LHV of anhydrous biomass) would be considerably improved (Figure 2.24).

2.3.5. *Methodology for modeling tar formation and conversion in a fluidized-bed reactor*

An approach is currently being developed in our group with *Electricité de France* (EDF) to model tar formation and conversion in the air-blown fluidized bed. It is presented in Figure 2.25. The main issue to overcome is the coupling among the hydrodynamic, heat and mass transfer and detailed chemistry, which are required for modeling tar.

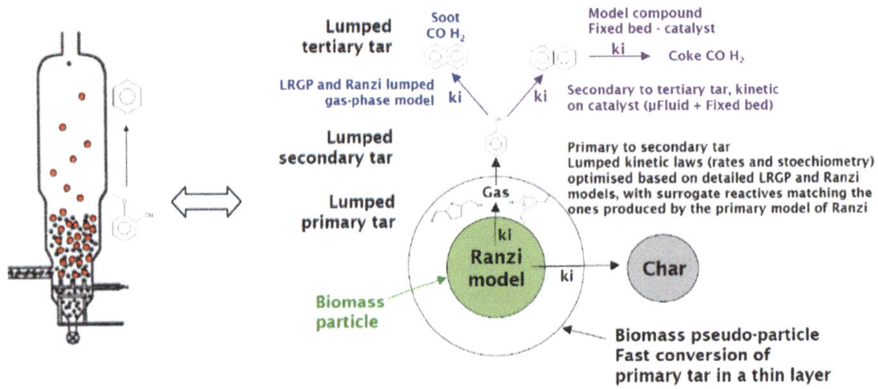

Figure 2.25. *Methodology currently developed at CNRS-Nancy (in collaboration with EDF) for modeling tar formation and conversion in gasification fluidized beds. For a color version of the figure, see www.iste.co.uk/dufour/biomass.zip*

In this approach, a particle model has been developed (by Dr. G. Mauviel and Dr. O. Authier), which combines heat transfer and detailed primary pyrolysis kinetic based on the model developed by Ranzi *et al.* [RAN 08]. The heat and mass transfer and the hydrodynamics in the reactor are also modeled. The biomass particle generates primary tars that are very rapidly decomposed into secondary products at 850°C in the fluidized bed. It seems very difficult to analyze these secondary products during their formation. The yield and composition of lumped secondary products are predicted based on the modeling under the Chemkin® IV software of Ranzi *et al.* [RAN 08] and LRGP-CNRS (Nancy) gas-phase detailed models (thousands of elementary reactions). A lumped kinetic mechanism (surrogates, reactions and kinetic laws) is also derived from the detailed models for secondary tar conversion to tertiary tar at gas phase. The catalytic conversion (over dolomite or olivine, typical bed/catalyst for fluidized bed) of secondary and tertiary tars is also modeled by simplified kinetic laws determined from the experiments by our group on tar conversion over the catalysts (olivine and dolomite).

2.4. Reactors for biomass pyrolysis

2.4.1. *Wood torrefaction*

This technology has gained recently much interest, especially for the pretreatment of biomass for large-scale combustion or gasification units.

Torrefaction is slow pyrolysis (heating of biomass under inert atmosphere) at a low temperature (usually lower than 250°C). This treatment decreases the grinding energy required for producing small particles from biomass and produces a hydrophobic material that is more convenient to be transport or stored (much lower moisture content than wood).

An interesting work of *École des Mines* in Saint Etienne (France) has shown that there is a trade-off to find between the mass loss and the grinding energy. They have defined a "grinding criterion" equal to: grinding energy (kWh/ton of biomass)/ volumic % of obtained small particles (<200 μm). This criterion does not decrease significantly after a mass loss of only 10%. Thus, 10% mass loss during torrefaction may be a good optimum [REP 10].

2.4.2. *Char uses and production*

The various uses of char are presented in Figure 2.26 and ranked from the left to the right based on the added-value of the use.

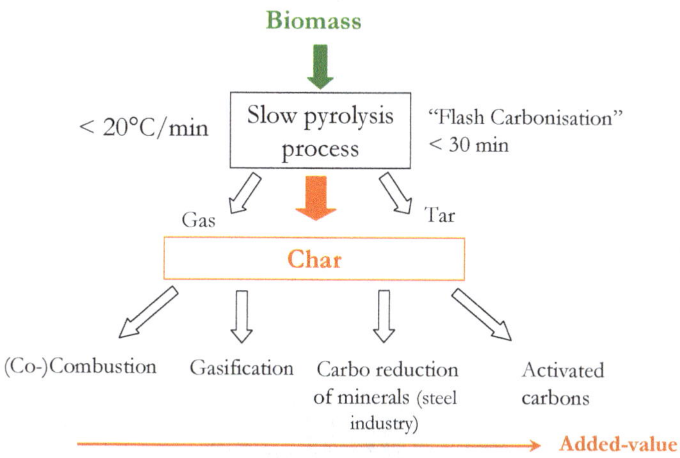

Figure 2.26. *Potential valorization routes of char. For a color version of the figure, see www.iste.co.uk/dufour/biomass.zip*

The lowest added-value can be combustion of char as in a cookstove. The pyrolysis step improves the energy density of bioenergy (higher energy density of char vs. wood; see the decrease in O content in the Van Krevelen diagram (Figure 2.30)) and reduces the emission of indoor pollutants. Indeed, pollutants from wood cookstove mainly come from incomplete oxidation of pyrolysis products. When the pyrolysis is conducted in an appropriate setup (outside the house!), the indoor emission from char combustion will be reduced.

Char can be gasified as wood. The pyrolysis step would reduce the transport cost and the heterogeneity in the composition of biomass by controlling the pyrolysis. Indeed, char could be mixed with coal or fossil fuels and could be a good intermediate to produce fuels and chemicals in large-scale gasification units (Figure 2.27).

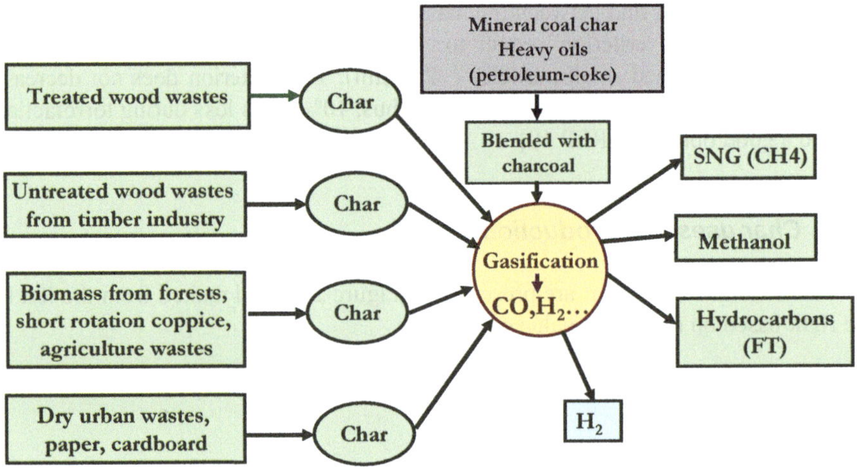

Figure 2.27. *Char as a potential intermediate for large-scale gasification and cogasification with fossil fuels [DEG 08]*

Char can be produced in traditional batch reactors or in continuous processes (Figure 2.28).

(a) (b)

Figure 2.28. *(a) Batch reactors for char production (here in Brazil [PER 09]) and (b) concepts for continuous process*

In continuous processes, heat can be produced by internal heating (moderate well-controlled oxidation of pyrolysis products) and by external heating with an external fuel source or with pyrolysis gas.

Old continuous processes were developed (in the middle of the 20th Century) and some of these technologies are presented in Figure 2.29.

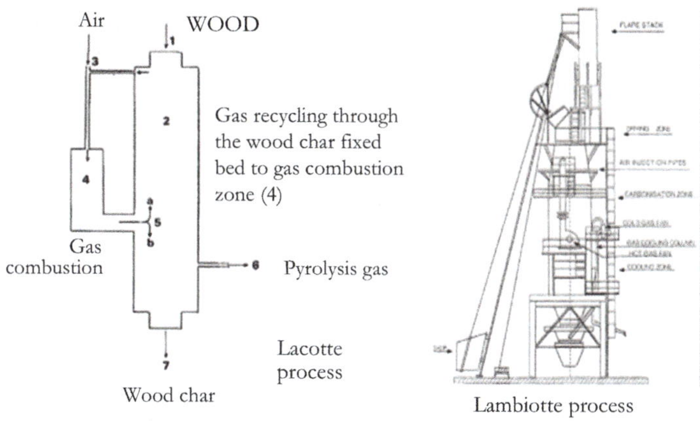

Figure 2.29. *Some old technologies developed for continuous wood "pyrolysis" or "carbonization" or "distillation"*

The mass yields and energy distribution of char depend on biomass type, furnace heating rate, pressure in the reactor, etc.

Slow heating rate and high pressures promote the formation of char [ANT 03].

Char mass yields for some pyrolysis conditions are given in Table 2.7.

Materials		Char yields (%)	Liquid yields (%)	Gas yields (% by diff)
Sawdust	1	37.00	45.11	17.89
	2	42.00	9.55	48.45
Sugar cane bagasse	1	36.90	46.10	17.00
	2	42.50	9.11	48.39
Olive oil cake	1	39.70	40.10	20.20
	2	43.00	9.00	48.00

Table 2.7. *Mass yields of char for various pyrolysis conditions [GUE 91]*

Mass yields and energy distribution obtained for a fixed bed, in batch mode with a furnace set at 500°C, are presented in Table 2.8 [RAV 96].

	Higher heating value				Energy distribution (%)		
	Biomass	Char	Liquids	Gas	Char	Liquids	Gas
Biomass							
Coir pith	19.5	25.0	18.7	16.1	37.9	28.3	33.9
Corn cob	16.1	28.6	23.8	5.2	35.6	49.4	15.0
Groundnut shell	19.8	27.4	23.6	10.0	37.5	44.7	17.9
Rice husk	20.0	44.2	22.5	7.4	38.3	46.2	15.5
Wood	20.0	24.1	24.9	16.6	23.4	28.3	48.3
De-ashed biomass							
Coir pith	–	26.2	22.3	9.9	42.1	41.4	16.6
Corn cob	–	26.4	24.2	5.0	21.2	65.2	13.7
Groundnut shell	–	29.8	21.8	6.2	41.2	50.5	8.3
Rice husk	–	31.0	19.5	6.6	37.9	56.1	6.0
Wood	–	24.2	28.5	11.3	16.5	57.3	26.2
Components							
Cellulose	11.7	32.4	16.2	1.3	30.8	64.6	4.5
Lignin	24.1	33.3	28.7	8.2	57.7	31.9	10.4
Xylan	30.4	32.3	22.5	37.6	22.0	30.0	48.0

Table 2.8. *Higher heating value (MJ/kg on dry ash free basis) of biomass and component pyrolysis products [RAV 96]*

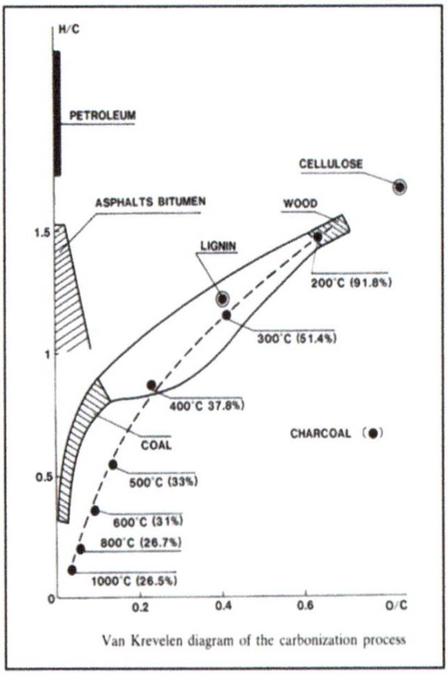

Van Krevelen diagram of the carbonization process

Figure 2.30. *The Van Krevelen diagram of the carbonization process [DEG 82]*

Char yields can be improved by high pressures in pyrolysis reactors [ANT 03].

The composition of char can be modified by controlling pyrolysis conditions (such as final temperature and pressure). The Van Krevelen diagram (Figure 2.30), which presents the atomic ratio H/C as a function of the atomic ratio O/C of various fuels, depicts the different compositions of char as a function of final temperature of pyrolysis.

2.4.3. *Pyrolysis bio-oils uses and production*

The main uses for fast pyrolysis bio-oils are presented in Figure 2.31.

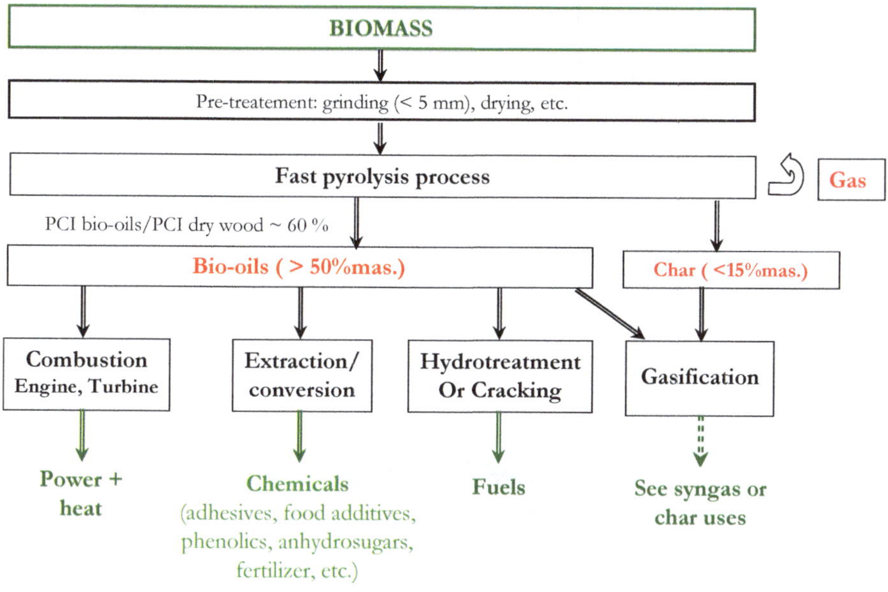

Figure 2.31. *Potential uses of bio-oil*

The potential applications of bio-oil have been extensively reviewed [CZE 04]. The main current commercial uses of fast pyrolysis bio-oils are food additives ("liquid smoke," Ensyn, Canada) or combustion (Metso, Finland). Bio-oils could also be a good intermediate for electricity production through diesel engine combustion [BRI 02]. Numerous works have dealt with bio-oil catalytic upgrading by hydrodeoxygenation [ELL 15a] or catalytic cracking (over zeolite) [VAL 12] to produce biofuels or green aromatic chemicals. Bio-oil could also be a good intermediate to reduce the transport cost of biomass for large-scale gasification or combustion units.

Figure 2.32 gives an overview of a potential biorefinery based on fast pyrolysis.

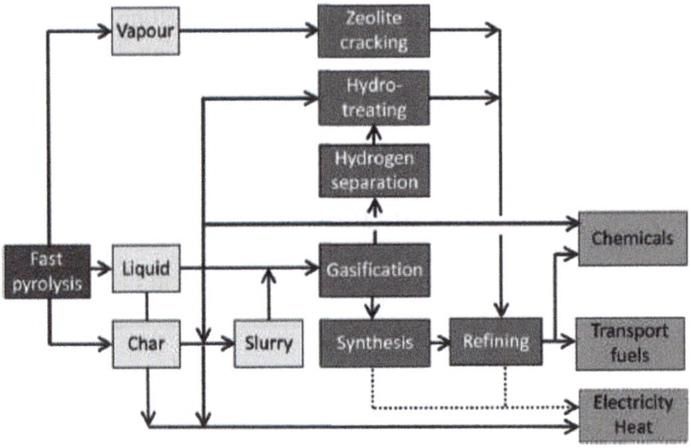

Figure 2.32. *Fast pyrolysis-based biorefinery [BRI 12]*

The chemical mechanisms of catalytic upgrading of bio-oils (by hydrotreatment or cracking) are discussed in Chapter 4.

Fast pyrolysis bio-oil is a dark and viscous liquid (Figure 2.33). Its density is about 1,200 kg/m^3.

Figure 2.33. *Fast pyrolysis bio-oils [VEN 10]*

The main differences between bio-oil and heavy fuel oil are presented in Table 2.9. The main issues to be tackled for a better valorization of bio-oils are high oxygen content (which results in high reactivity and corrosion issues during storage or handling and phase separation), particles, LHV, etc.

	Bio-oil	**Heavy fuel oil**	**Remark**
Water content (wt.%)	15–30	*0.1*	
pH	2.5		Acid
Density	1.2	*0.94*	
Composition (wt%)			
C	54–58	*85*	
H	5.5–7.0	*11*	
O	35–40	*1.0*	Not miscible in hydrocarbons
N	0–0.2	*0.3*	
S	<0.05	*2.5*	
Inorganics	0–0.2	*0.03*	Deactivation of catalysts
HHV (MJ/kg)	16–19	*40*	
Solid particles (wt.%)	0.2–1.0	*1*	Erosion of injector
Distillation residue (wt.%)	Up to 50	*1*	Instability and phase separation

Table 2.9. *Physical–chemical properties of wood pyrolysis bio-oil [CZE 04] adapted by Dr. G. Mauviel CNRS-Nancy*

The bio-oil is a complex mixture of hundreds of species, which requires a specific method for analysis. Figure 2.34 gives a fractionation scheme for the chemical characterization of bio-oil. Table 2.10 shows the high variety of chemical moieties present in bio-oils with mainly acids, alcohols, aldehydes, sugars and aromatics (from lignin).

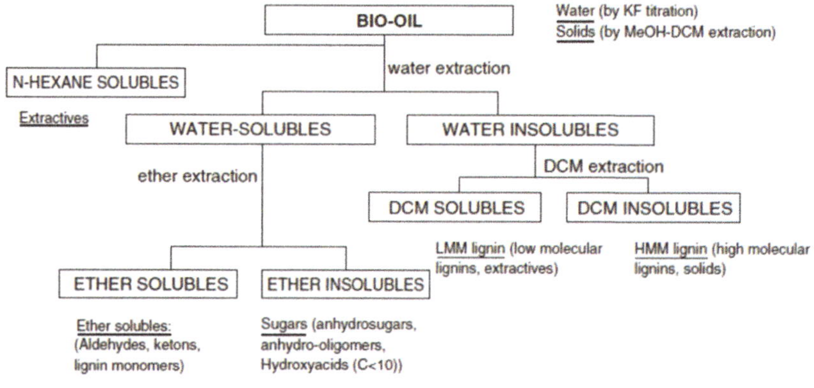

Figure 2.34. *Fractionation scheme for chemical analysis of bio-oil [OAS 03]*

Reference Pine Oil		Wet	Dry	C	H	N	O
Water	wt-%	**23.9**	0				
Acids	wt-%	**4.3**	**5.6**	40.0	6.7	0	53.3
Formic acid	wt-%		1.5				
Acetic acid	wt-%		3.4				
Propionic acid	wt-%		0.2				
Glycolic acid	wt-%		0.6				
Alcohols	wt-%	**2.23**	**2.9**	60.0	13.3	0	26.7
Ethylene glycol	wt-%		0.3				
Isopropanol	wt-%		2.6				
Aldehydes and ketones	wt-%	**15.41**	**20.3**	59.9	6.5	0.1	33.5
Nonaromatic aldehydes	wt-%		9.72				
Aromatic aldehydes	wt-%		0.009				
Nonaromatic ketones	wt-%		5.36				
Furans	wt-%		3.37				
Pyrans	wt-%		1.10				
Sugars	wt-%	**34.44**	**45.3**	44.1	6.6	0.1	49.2
Anhydro-β-D-arabino-furanose, 1,5	wt-%		0.27				
Anhydro-β-D-glucopyranose (Levoglucosan)	wt-%		4.01				
Dianhydro-α-D-glucopyranose, 1,4:3,6	wt-%		0.17				
LMM lignin	wt-%	**13.44**	**17.7**	68	6.7	0.1	25.2
Catechols	wt-%		0.06				
Lignin derived Phenols	wt-%		0.09				
Guaiacols (Methoxy phenols)	wt-%		3.82				
HMM lignin	wt-%	**1.950**	**2.6**	63.5	5.9	0.3	30.3
Extractives	wt-%	**4.35**	**4.35**	75.4	9.0	0.2	15.4

Table 2.10. *Chemical composition of reference pine fast pyrolysis oil and its fraction [OAS 03]*

The selectivity in some compounds (e.g. levoglucosan) in bio-oils can be increased, for instance by catalytic upgrading or by acid washing and acid impregnation of biomass. Levoglucosan could be an interesting intermediate for further biological conversion (e.g. fermentation). Acid washing removes ashes from biomass and enhances the yield in levoglucosan [PEC 15]. Indeed, ashes in biomass inhibit the formation of levoglucosan and bio-oils to promote the formation of char.

The production of bio-oils is promoted by a fast heating rate of biomass or rather by a high heat flux density (W/m^2, power by unit surface of particles) brought by the reactor to the biomass [LÉD 10, LÉD 15]. A gas-phase temperature in the reactor no higher than about 500°C is also required. The primary tar formed during fast pyrolysis should be rapidly quenched to have a low residence time in the hot zone of the reactors (less than 2 s at 500°C). They are condensed and then result in the production of "bio-oils."

Fast pyrolysis technologies are presented in Figure 2.35. They have been comprehensively reviewed in many studies [CZE 04, VEN 10, BRI 12].

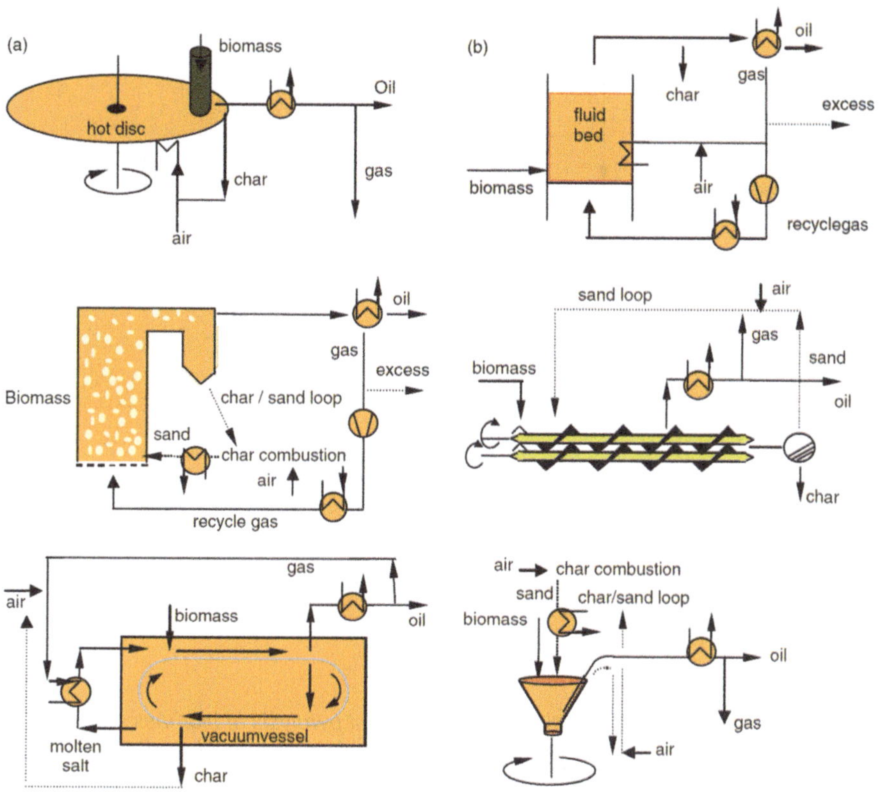

Figure 2.35. *Various technologies for biomass fast pyrolysis: (a) ablative, circulating fluidized bed and vacuum technologies, and (b) dense fluidized bed, screw (Auger) reactor and rotating cone [VEN 10]*

In all these technologies, biomass is heated by a solid heat carrier (sand in Auger or fluidized-bed reactor, hot disk or cone) or a liquid heat carrier (molten salt). The good contact between biomass and solid or liquid heat carriers promotes heat transfers from the reactor to the surface of particles and, thus, increases the heating rate. The main issues are the separation of primary vapors with chars (e.g. by cyclones) and then efficient quenching (by scrubbers) of vapors to produce bio-oils.

An original technology has been developed in our laboratory in the early 1980s: the cyclone fast pyrolysis reactor (Figure 2.36). The properties of bio-oils produced by this reactor have been comprehensively discussed [LÉD 07].

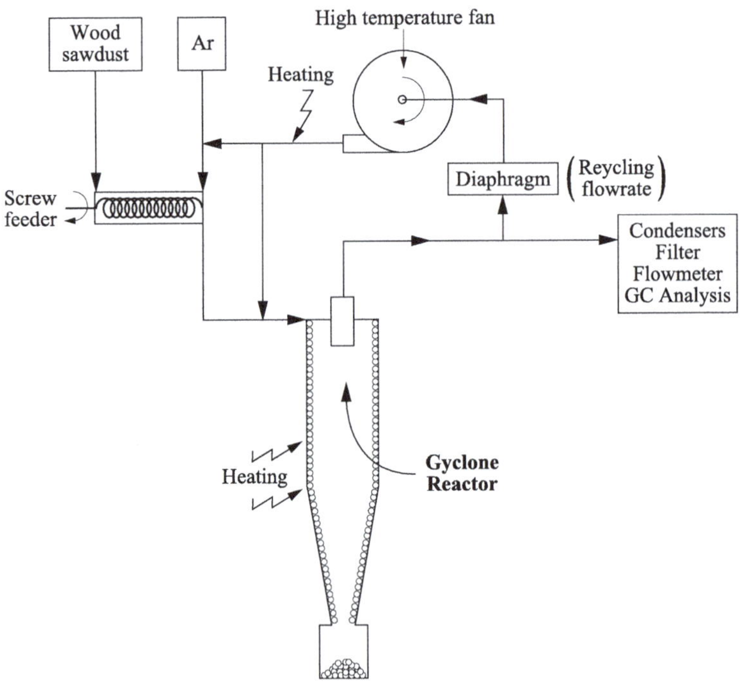

Figure 2.36. *The cyclone reactor for fast pyrolysis of biomass: an original, compact and multifunctional reactor developed in CNRS-Nancy in 1980s [LÉD 86]*

It is a very compact reactor that is multifunctional because it acts as a heating device and a separator of char and vapors [LÉD 86].

Figure 2.37 gives the mass and energy balance of fast pyrolysis process based on a fluidized-bed reactor. It is important to notice the very high flow rate of sand (66,600 kg/h, which is about 30 times the flow rate of biomass), which is recycled in order to bring the required heat for pyrolysis.

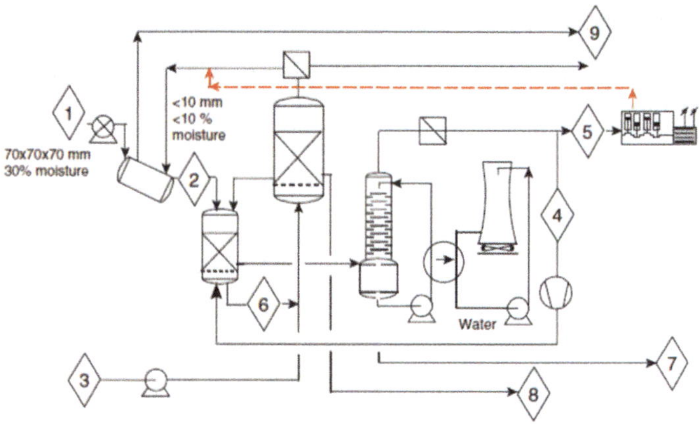

	Input	Reactor feed	Air	Inert carrier gas	Pyrolysis vapours	Heat carrier	Bio-oil	Surplus char	Flue gas
Stream no.	1	2	3	4	5	6	7	8	9
Organics	2,000	2,000					1200		
Water	900	200					400		740
O_2			600						250
N_2			2,400						2,400
CO_2									470
Pyrolysis gas				1,000	200				
Char						155		245	
Ash	20	20						12	8
Sand						66,600			
Total	2,920	2,220	3,000	1,000			1,200	257	3,860
Temp. (K)	291	291	291	326	304	804	317	804	400
Pressure (bar)	1	1	1	1.2	1	1	1	1	1
Chem. heat (MW)	10.4	10.4		2.1	0.42	1.3	6.6	2.1	
Heat (MW)						7.41		0.05	0.33

Figure 2.37. *Mass and energy balance for 2,000 kg/h (biomass, dry ash free basis) fluidized-bed pyrolysis process [VEN 10]*

2.5. Reactors for biomass liquefaction

Herein, by the term "liquefaction" we mean all the processes in which biomass is converted into a liquid medium at high pressures. The early research work on biomass liquefaction has been reviewed by Elliott *et al.* [ELL 91]. More recently, this process has been paid an increasing interest in the frame of lignin valorization for the production of green chemicals (it can be named as "catalytic depolymerization") [ZAK 10].

This technology consists of heating biomass (or one of its components such as lignin) in the presence of a solvent (water, tar, ethanol, etc.), a reductive gas (e.g. H_2) and catalysts under high pressure to maintain the solvent at the liquid phase. The goal is to produce bio-oils with improved properties than fast pyrolysis bio-oils.

Liquefaction and fast pyrolysis are compared in Table 2.11.

Advantages of liquefaction vs. fast pyrolysis	**Drawbacks of liquefaction vs. fast pyrolysis**
Biomass with higher variability can be treated: not limited by water content and less limited by particle size Possible coliquefaction with coal, wastes, etc. High conversion yield in bio-oils to 60%mas. dry wood with low oxygen fraction (<10%mas.) than bio-oils from fast pyrolysis (~30%mas.) [BRI 88] Lower yield in char (<5%mas.) (for a given particle size) than fast pyrolysis (~10%, function of experimental conditions)	Technological problems due to high H_2 pressure (50–200 bar) (but conditions closed to hydrotreatment conditions in petroleum refineries or coal liquefaction) High viscosity of oil from liquefaction Lack of demonstration plants vs. fast pyrolysis [ELL 91] Deactivation and poisoning of catalysts Inorganic content could be higher in liquefaction bio-oil

Table 2.11. *Comparison between high-pressure liquefaction and fast pyrolysis for bio-oil production*

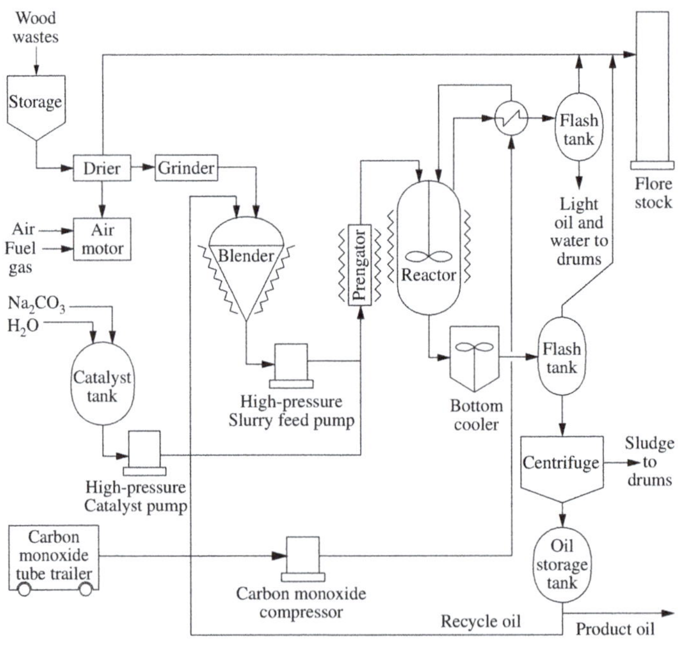

Figure 2.38. *Pilot plant for biomass liquefaction developed in Albany in early 1970s [LIN 78]*

The early development of this technology started in 1960s and 1970s with the Noguchi process [GOH 66] and with a pilot plant operated in Albany [ELL 91], which is presented in Figure 2.38.

Some problems for this technology are the following:

– wood should be grinded at a low particle size to prepare a slurry;

–this slurry should be fed to a high-pressure reactor (e.g. plug flow tubular reactor or ebullated bed);

– the products may be diluted in the solvent, which implies recovery issues;

– the catalysts may be deactivated fast by the slurry (and heavy products, coke, etc.) and should be recovered and regenerated/recycled (and/or very cheap), etc.

A mixture containing as much as 60% wood flour in product oil has been pumped into pressurized containers using a modified extruder as a feeder at the University of Arizona [ELL 91]. Technological developments could also come from the crude oil or coal hydroprocessing industry.

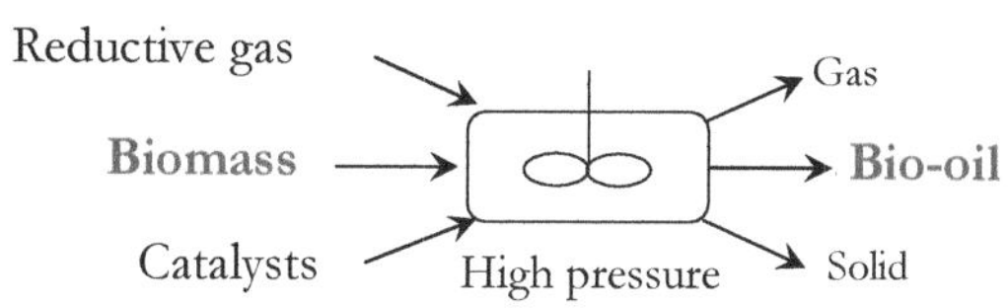

Source	Solvent	Catalysts / co-reactive	Reduction gas	Pressure bar	Temp. °C	Yield (%mas. oil /dry wood)
Van Krevelen, 1958	water	Iode	H_2	440	325	40
Appell, 1977	water/cresol	HCOONa/ HCOOH	none	72	250	41
Elliott, 1977	Anthracene	Na_2CO_3	CO	102	350	65
Boocock, 1979	water	Ni Raney	H_2	200	350	37
Kafrouni, 1982	water	Na_2CO_3	CO	15 to ~150	300	45
Meier, 1987	Wood tar	Co, Mo, Cr, Ni, Pd, Fe, NiMo, etc.	H_2	30-130	380	26-37
Wang, 2007	none, toluene, tetraline	Mo	H_2 or syngas	20	300-350	30-65

Table 2.12. *Some operating conditions tested for biomass liquefaction*

An overview (not exhaustive) of some operating conditions is presented in Table 2.12. Experiments are conducted in stirred tank reactors at the laboratory scale but at the pilot scale, and to improve the selectivity in products, plug flow reactors would be a preferred configuration [SHA 03].

The as-produced bio-oils could be a good "green" solvent for biomass liquefaction [GOH 66]. More recently, it has been shown that liquefaction (in guaiacol as a solvent) produces heavy compounds, which are in too high a yield to allow an effective recycling of the bio-oil, as an internal solvent [KUM 14].

A carbon balance is presented in Figure 2.39 for the hydrothermal liquefaction. This process is meant to treat wet biomass without drying and to access ionic reaction conditions by maintaining a liquid water processing medium [ELL 15b].

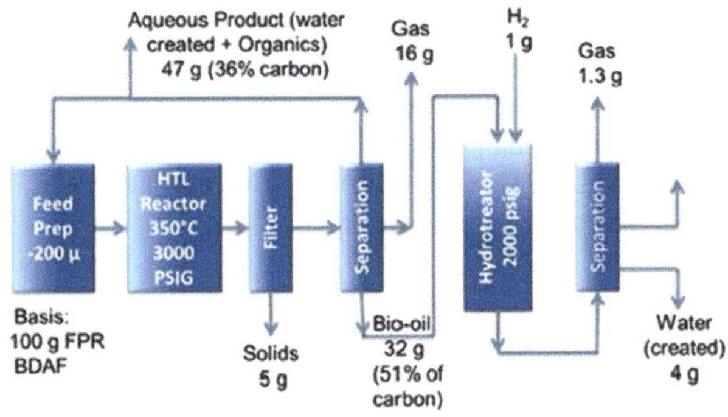

Figure 2.39. *Carbon balance for hydrothermal liquefaction of biomass to fuels*

3

Particle Scale and Mesoscale

3.1. Introduction on the main mechanisms occurring during wood pyrolysis

We have previously mentioned that pyrolysis is the first step occurring in all thermochemical reactors.

Figure 3.1 depicts the main physical and chemical phenomena occurring during the pyrolysis of a biomass particle.

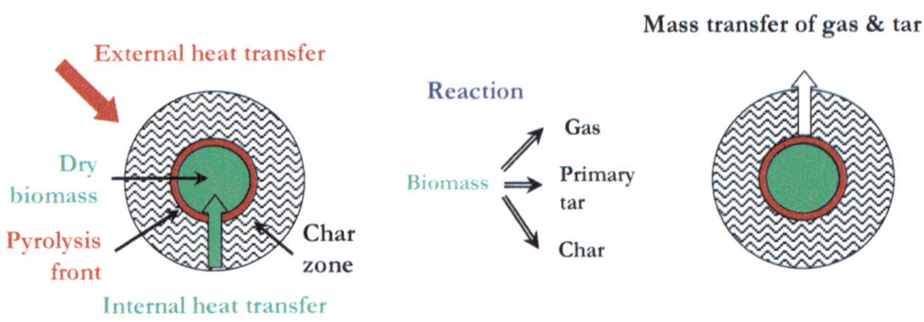

Figure 3.1. *Heat transfers, reaction (chemical kinetic) and mass transfers occurring during the pyrolysis of a biomass particle. For a color version of the figure, see www.iste.co.uk/dufour/biomass.zip*

In all thermochemical reactors, particles are heated: this is characterized by an external heat transfer coefficient (W/m^{2}K). During heating, the biomass particle then undergoes decomposition to produce volatiles and char. An important thermal gradient is produced inside the particle, especially when a high heat flux density

(W/m^2) is brought as in fluidized bed. Volatiles or liquid droplets are then ejected from the pore structures of the biomass/char particle to the surrounding gas. Figure 3.2 highlights these main phenomena and their multiscale character. The particle size and structure have been shown to considerably affect the composition of products (mainly of tar or bio-oil) because of mass transfer limitations in the cell walls and the network of pores [GAS 11, ZHO 14].

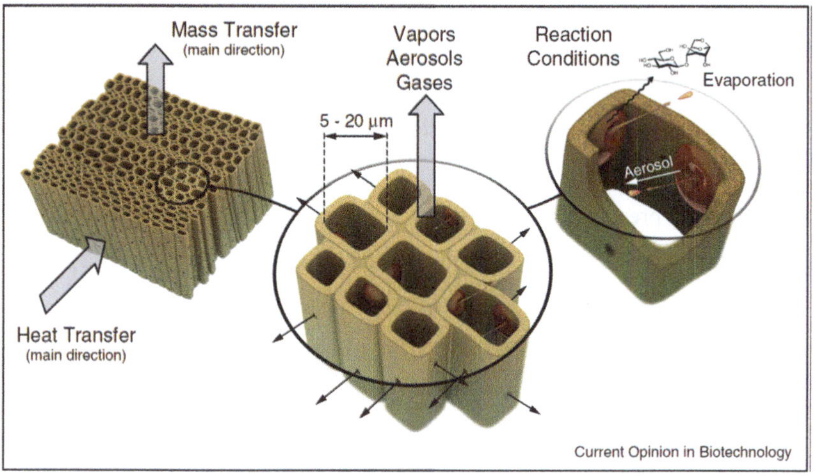

Figure 3.2. *Schematic of the multilevel character of biomass pyrolysis (for softwood) ranging from a single cavity (cells) to a wood particle [KER 13]*

3.2. Physical–chemical mechanism of pyrolysis at the particle scale

3.2.1. *Main characteristic times of phenomena occurring during pyrolysis*

A comprehensive mathematical model of complex chemical and physical processes inevitably involves assumptions and approximations, which can be based on the analysis of characteristic time scales. This kind of approach has been used in the development of many pyrolysis models [CHA 85, MOG 06, AUT 09]. Timescales do not reflect the magnitude of the driving force (such as pressure for gas convection) but do indicate the rate-controlling phenomena. The main characteristic times associated with the pyrolysis of biomass [AUT 09] are:

internal heat transfer by conduction: $t_{hc} = \dfrac{\rho C_p L^2}{\lambda}$

internal mass transfer by diffusion: $t_{dm} = \dfrac{L^2}{D_f}$

internal mass transfer by convection: $t_{cm} = \dfrac{L^2 \mu_f}{P_k K_f}$

external heat transfer: $t_{ht} = \dfrac{\rho C_p L}{h}$

pyrolisis time for a 3-parallel reaction scheme: $t_p = \dfrac{1}{k_P} = \dfrac{1}{k_G + k_V + k_C}$

The variables presented in the earlier-mentioned equations correspond to:

– L (particle dimension) (for instance = 0.003 m);

– ρC_p (mass density (kg/m^3) × heat capacity (J/kg K)) ~ 10^5 J/m^{3}K;

– λ (thermal conductivity) ~ 0.10 W/m K, for char [GRO 00];

– K_f (wood permeability) ~ 10^{-14} m^2;

– μ_f (gas viscosity) ~ 10^{-5} Pas;

– P_f (volatile overpressure inside the particle) ~ 10^5 Pa;

– D_f (gas diffusivity inside the wood char) ~ 10^{-6} m^2/s;

– k_p (kinetic constant for pyrolysis) = 10^{-9} exp(–120 (kJ/mol)/RT) (s^{-1}) [GRA 84];

– h (external heat transfer coefficient) = 500 W/m^{2}K (for instance, in a fluidized bed).

The characteristic times can be compared by the following dimensionless numbers:

Biot number: $Bi = \dfrac{hL}{\lambda} = \dfrac{t_{hc}}{t_{ht}}$

Lewis number: $Le = \dfrac{\mu_f \lambda}{P_f K_f \rho C_p} = \dfrac{t_{cm}}{t_{hc}}$

Peclet number: $Pe = \dfrac{P_f K_f}{D_k \mu_f} = \dfrac{t_{dm}}{t_{cm}}$

Thiele number: $Th = \dfrac{\rho C_p L^2 k_p}{\lambda} = \dfrac{t_{hc}}{t_p}$

characteristic times (s)	wood	char	T_{ref} (K)	ϕ_i (MW·m^{-2})
t_{hc}	23.4	21.8	600	
t_{dm}	9.0			
t_{cm}	9.0×10^{-2}	9.0×10^{-4}		
t_{ht}	4.6	1.2	600	0.8
	–	0.7	1800	0.8
	8.6	2.3	600	0.3
	–	1.0	1800	0.3
t_p	1.6	–	800	
	2.9×10^{-2}	–	1000	
dimensionless number				**orders of magnitude**
Pe				~10^2–10^4
Le				~10^{-5}–10^{-3}
Bi				~1–10
Th				~1–10^3

Table 3.1. *Characteristic times and dimensionless numbers of main processes involved during fast pyrolysis of wood [AUT 09]*

In the conditions investigated by Authier *et al.* [AUT 09], the characteristic times and dimensionless numbers are given in Table 3.1.

In the conditions investigated by Authier *et al.* (Table 3.1), Peclet number (Pe) is much greater than 1, showing that volatile species transfer by diffusion is slower than that by convection. As a consequence, a model taking into consideration volatiles release could disregard diffusion in the gas mixture phase comparative with the convective mass transfer.

Lewis number (Le) is much less than 1 (Table 3.1), showing that the pyrolysis model should mainly focus on internal heat transfer rather than on mass transfer.

Biot number (Bi) is greater than 1. Therefore, internal heat transfer by conduction is slower than external heat transfer. Because of intraparticle thermal gradients, the temperature distribution in the whole particle volume is not uniform.

Finally, Thiele number (Th) is greater than 1, showing that internal heat transfer proceeds slower than the pyrolysis reaction. Consequently, pyrolysis reactions occur inside a thin zone, which moves from the surface of the sample toward the center of the particle (see Figure 3.1).

3.2.2. *Modeling intraparticle mechanisms of biomass pyrolysis*

It has been known for at least one century that biomass forms an intermediate "liquid-like" phase or "molten residue" during pyrolysis [GOR 63, LÉD 87, KLA 10] even if it has been controversial [NAR 96, MAM 09]. Some researchers did not notice the molten residue at a slow heating rate maybe because the observation of particles was not conducted at the suitable scale needed to see the molten material.

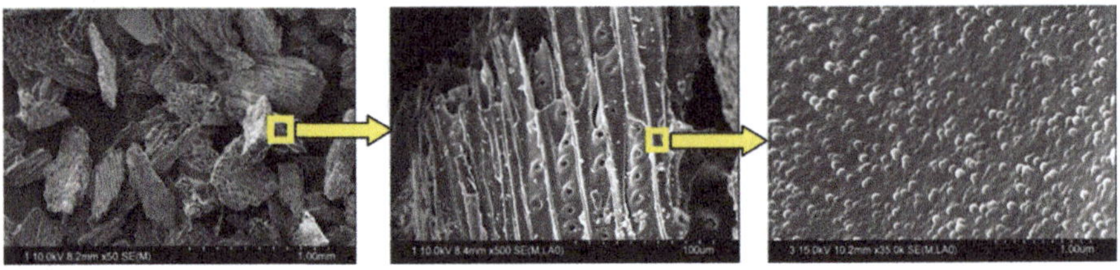

Figure 3.3. *Scanning electron microscope (SEM) pictures of a wood char produced under a slow heating rate from millimeter to micrometer length scales. The overall wood cell macrostructure is preserved at the 0.1-mm scale despite the important loss of cell wall material after pyrolysis [HAA 09]. Bubbles on the cell wall are formed by volatiles pushing a viscoelastic material at the micrometer length scale [FIS 02]*

Figure 3.3 was obtained by our group from a wood char pyrolyzed under slow heating rate profile [DUF 11b]. It shows that, under these thermal conditions, the overall wood cell macrostructure is preserved at the 0.1-mm scale despite the important loss of cell wall material after pyrolysis, in agreement with Haas *et al.* [HAA 09]. Nevertheless, the formation of bubbles by volatiles pushing a viscoelastic material can be observed at the cell wall surface, at the micrometer scale.

On the basis of this observation, the characteristic times considered in one study of our group [DUF 11b] were the following:

– chemical kinetics;

– liquid diffusion in a liquid phase (tar at liquid phase);

– liquid convection in a porous solid;

– gas diffusion in a porous solid;

– gas convection inside the particle;

– internal heat transfer that is supposed to occur mainly through conduction although internal radiation can also be important [CHA 85].

External mass and heat transfers were not considered in this study.

The effect of the reference temperature and the length scale on the characteristic times has been discussed [DUF 11b]. An important consideration for the analysis of characteristic times is the relevant length scales, which range from:

– 1 nm, the molecular scale, at which the reactions occur;

– to 1 μm, average size of bubbles depending of the heating rate and of many other conditions;

– 10 μm, diameter of tracheids in wood and of macropores of char, where most of the porous volume is located [EHR 82];

– 1 mm–1 cm, the characteristic length of a particle.

The characteristic time of transfer phenomena depends on the square of the reference length scale.

At the molecular scale (<1 nm), the reaction is most probably the limiting phenomena. Volatiles can then diffuse in their liquid phase into bubbles or to the pore surface where evaporation occurs (or to the particle surface if the particle is mainly in a liquid phase). Characteristic times of liquid diffusion as a function of characteristic length (1.2×10^4 s for 1.5 mm, 5.5×10^{-3} s for 1 μm, at 700 K) show that volatiles formed within the melt are much more likely to diffuse (in their liquid phase) into bubbles rather than to the particle surface. The characteristic time of diffusion to the particle surface (1.2×10^4 s for 1.5 mm) is much too high to explain the observed pyrolysis rates (about 30 s for a particle of 1.5-mm thickness with 40 W/m^2K external heat transfer coefficient). Nevertheless, liquid (or condensed adsorbed phase) diffusion to the pores' surface where evaporation occurs could be a limiting step for smaller particles and high external heat transfer coefficient as in fluidized beds.

Bubbles should constitute a major vehicle for intraparticle transport of the liquid phase as experimental evidence of bubble formation during biomass pyrolysis was obtained through scanning electron microscope (SEM) analysis. The characteristic time for the transport of volatiles through bubbles is difficult to estimate [OH 89].

Permeability depends, among other parameters, on the length scale, the conversion degree and the presence of the viscous phase. Once the viscous phase is developed, the viscous flow can close off the pores [OH 89], thus leading to an important decrease in permeability. The range of wood permeabilities is five orders of magnitude. The permeability of char is three orders of magnitude higher than that of wood, and the permeability of the liquid phase would be much smaller than those of wood and char. The characteristic time of gas or liquid convection could be very high if the permeability of metaplast is considered, thus leading to a high gas and liquid accumulation in the pores as shown by Haas *et al.* [HAA 09].

Given the uncertainty of permeability and the fact that gas convection could possibly be of great importance as a faster mass transfer phenomena than diffusion, the model we developed considered gas convection. Moreover, given that there is limited understanding of the conditions of the presence of a liquid phase and its transfer phenomena (probably through bubbles), the model handles the formation and conversion of this intermediate liquid phase (or "metaplast" as it is called for coal) [DUF 11].

The model (depicted in Figure 3.4) accounts for a simplified multistep chemical decomposition with the formation of liquid-phase tars inside the particle. The intermediate solid can be associated with the formation of activated cellulose (cellulose with a low degree of polymerization [BRO 73]). The liquid-phase tars are then competitively converted into a secondary char and gases or evaporated following the Clausius–Clapeyron law. To our knowledge, a tar evaporation law had so far never been coupled with biomass pyrolysis kinetics. The convective mass transport of all the volatile species through the porous particle is modeled by Darcy's law. This model offers a first approach to simulate the tar (at the liquid phase) lifetime and its intraparticle conversion. The predictions were compared with results from the literature [MIL 96] (Figure 3.5).

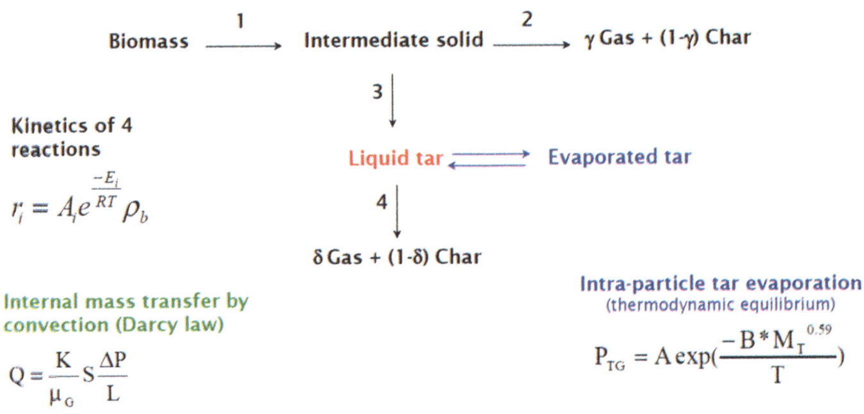

Figure 3.4. *Model developed for intraparticle mechanism of biomass pyrolysis [DUF 11]*

The model can still be considerably improved, for instance by including more detailed kinetics (e.g. from [RAN 08]) and a better modeling of heat transfers as developed in our group by Mauviel *et al.* [AUT 09].

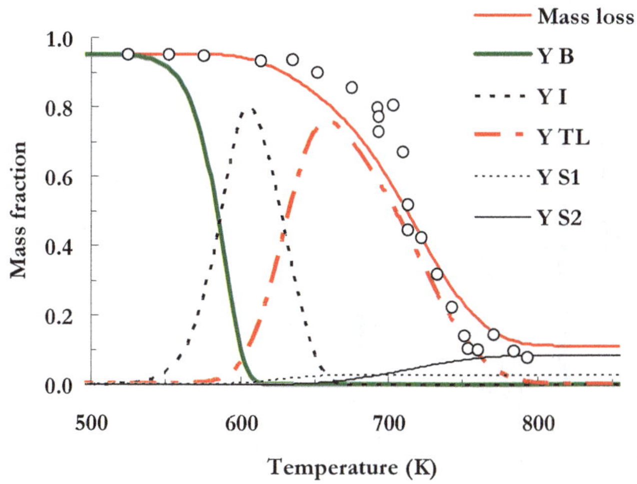

Figure 3.5. *Comparison between model predictions and experimental mass loss, taken from [MIL 96], for cellulose pyrolysis [DUF 11b]. YB, YI, YTL, YS1, YS2 represent mass fractions of cellulose, intermediate solid, liquid, char from intermediate solid and char from liquid, respectively. For a color version of the figure, see www.iste.co.uk/dufour/biomass.zip*

The Clausius–Clapeyron evaporation parameters were reviewed in this work and modified if levoglucosan or cellobiosan are assumed as the main tar compounds at liquid phase. The effects of these parameters on cellulose pyrolysis mass loss rate were modeled and discussed. The model shows that the limiting step for cellulose mass loss could be the conversion of the intermediate liquid phase (Figure 3.5). It also points out that mass transfer limitation can lead to a high intraparticle ovepressure and can control the lifetime of tar at liquid phase and the subsequent "secondary" char formation (S2 in Figure 3.5) from the conversion of liquid tar.

3.2.3. *Heat of biomass pyrolysis studied by calorimetry*

The approach presented in section 3.2.2. shows that mass transfer could considerably affect biomass pyrolysis. Our model [DUF 11b] does not yet account for the effect of mass transfer on the heat of pyrolysis reactions, which is a well-known fact [MOK 83, MIL 96]. Indeed, it has been shown that calorimetry is a very good analytical tool to investigate the effect of mass transfer on biomass pyrolysis and to further understand the chemistry of cellulose pyrolysis [MOK 83].

For this reason, our group has used a specific Calvet-type differential scanning calorimetry (DSC) (Setaram, France). It is composed of hundreds of thermocouples (fluxmetric detector) located around the two cylindrical zones (reference and sample) (presented in Figure 3.6).

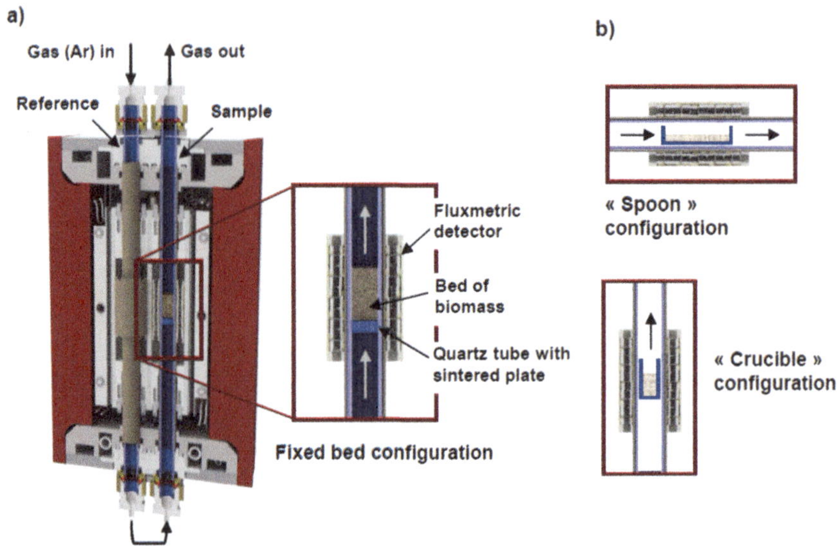

Figure 3.6. *(a) Scheme of a Calvet-type DSC cross section in the vertical mode under the quartz fixed-bed configuration (adapted from Setaram, France) and (b) simplified representation of "spoon" (horizontal) and "crucible" configurations*

It gives a quantitative integration of all the heat flux emitted by the sample, thus leading to a very high sensitivity, compared with other DSC probes based on only tens of thermocouples on a plate (instead of a cylinder). Because of this unique three-dimensional design, different geometries can be tested: (1) the crucible (platinum, vertical position of the calorimeter), (2) the "spoon" (platinum, horizontal position) and (3) the fixed bed (quartz tube with a sintered plate, vertical position) (Figure 3.6). To the best of our knowledge, this work presents the first comparison of these geometries and the first use of a fixed bed in a calorimeter to investigate the heat of biomass pyrolysis. The fixed bed allows improving and controlling the mass transfer from the bed of particles to the sweeping gas (Argon). Mass transfer was previously studied in closed or open pan [RAT 03] or under different pressures and flow rates [MOK 83] but not by using a fixed bed. The heating rate for the DSC was set to 10 K/min until 700°C.

The heat flows obtained for different carbohydrates, lignin and biomasses are presented in Figure 3.7.

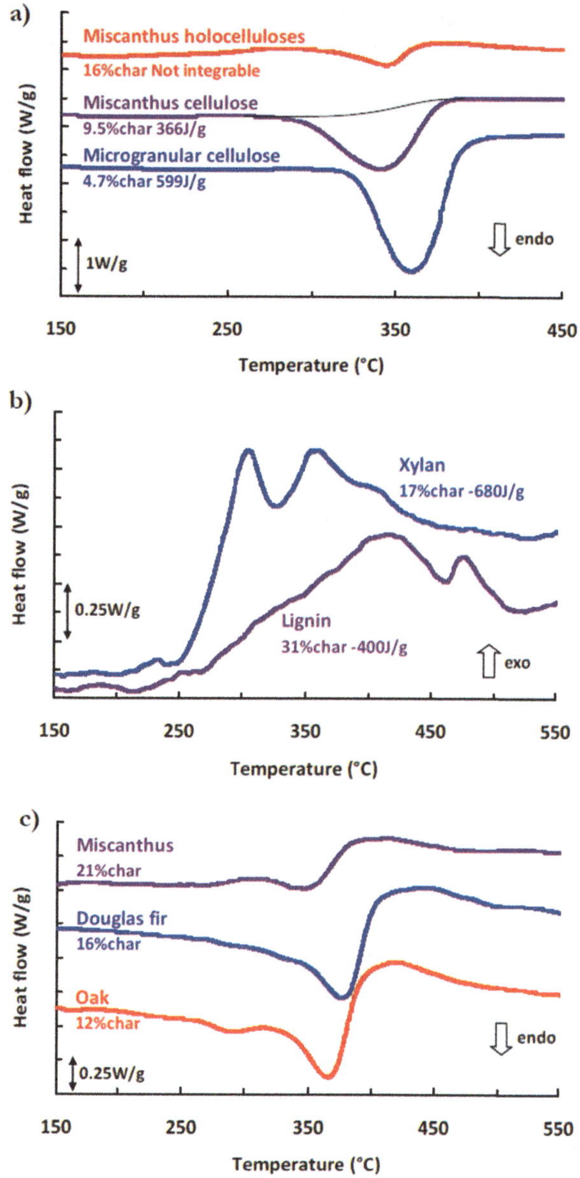

Figure 3.7. *DSC curves for (a) microgranular cellulose, Miscanthus cellulose and holocelluloses (20 mg, 20 mL/min fixed bed), (b) xylan and organosolv lignin (10 mg, 20 mL/min, spoon) and (c) Miscanthus, Douglas fir and oak (20 mg, 20 mL/min fixed bed) at 10 K/min. Char yield after 700°C final temperature. For a color version of the figure, see www.iste.co.uk/dufour/biomass.zip*

Different celluloses (crystallinity and mineral matter) and the effects of geometry, mass and flow rate of carrier gas on the heat of cellulose pyrolysis were studied (Figure 3.8).

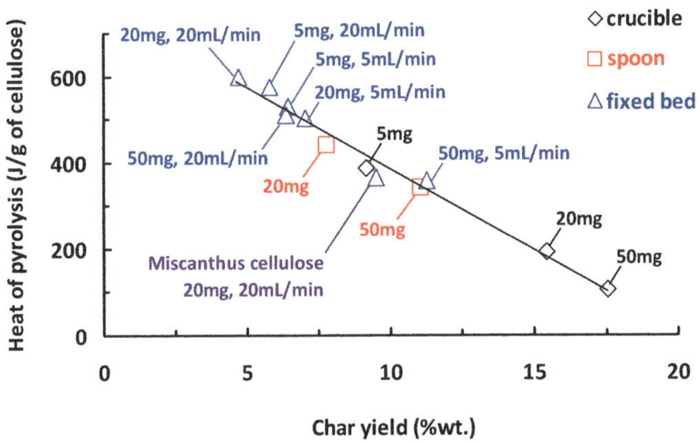

Figure 3.8. *Effect of geometry, mass, flow rate and type of celluloses on the heat of pyrolysis as a function of the char yield (all points for microgranular cellulose except the annotated point for Miscanthus cellulose)*

Cellulose pyrolysis forms intermediate liquid species (depolymerization and endothermic mechanism). Their evaporation (endothermic) or competitive conversion into char (crosslinking and exothermic) is very sensitive to the mass transfer mechanism [MOK 83, MIL 96, SUU 96]. This mechanism is depicted in Figure 3.9.

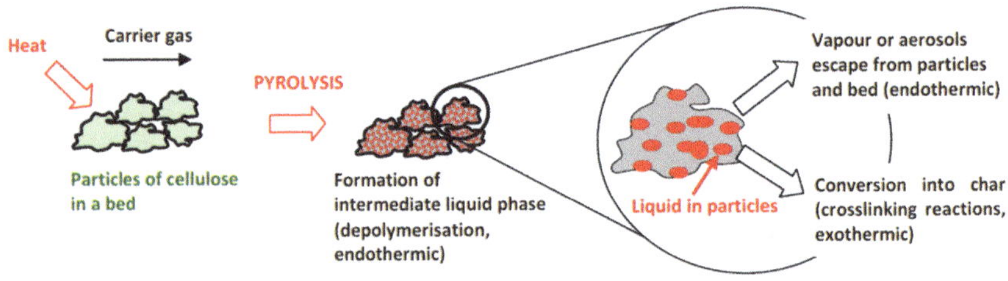

Figure 3.9. *Effect of mass transfer mechanisms on the heat of cellulose pyrolysis*

The fixed bed is the best design to reduce the mass transfer limitations (higher heat of pyrolysis and lower char yield, Figure 3.8). The slope of the curve obtained in Figure 3.8 (38 J/g cellulose per 1% increase of char yield) is in very good agreement with results from Mok and Antal [MOK 83] (35 J/g per 1% increase of char yield). The results of Mok and Antal were obtained under higher mass transfer limitations with spoon configuration and high pressure yielding higher char and enthalpy yields (10–22% char yield) than those in our case. This finding shows that the heat of cellulose pyrolysis versus char yield would lie on a straight line over a wide range of (slow) pyrolysis conditions. Unfortunately, calorimetry is not adapted to study fast pyrolysis because of the heat transfer characteristic time of the sensor. The slope of this straight line does not seem to depend on the cellulose composition (Figure 3.8). More experiments on various celluloses (degree of polymerization, crystallinity and ash content) are needed to confirm this finding.

Xylan and organosolv lignin pyrolysis mainly lead to exothermic signals (Figure 3.7(b)) because the enthalpy of bond breakage may be weak compared with the enthalpy of bond formation during crosslinking and the intermediate pool of species (and radicals) formed may promote crosslinking reactions (high char yield).

In our opinion, the DSC curves of holocelluloses, *Miscanthus*, oak and Douglas fir should not be integrated because of overlaps between endo- and exothermic phenomena. The thermal runaway during biomass torrefaction can be mainly attributed to xylan pyrolysis, which also leads to intermediate liquids with evaporation or crosslinking into char (exothermal) highly dependent on mass transfer.

3.3. Evolution of polymer network in biomass at the mesoscale

In situ ^{1}H NMR and rheology analysis have been extensively used for studying "metaplast" formation during coal pyrolysis [MIY 79, SAK 87, DÍA 05]. To the best of our knowledge, in situ ^{1}H NMR and rheology had never been used for biomass. Our group has used this methodology on biomass and its constituents (cellulose, lignin and xylan) [DUF 12a, DUF 12b] in collaboration with Colin Snape (the University of Nottingham) [STE 04, DÍA 08].

3.3.1. *In situ analysis by ^{1}H NMR*

The fundamentals of in situ ^{1}H NMR analysis are briefly recalled. This technique measures the protons transverse relaxation signal stimulated by a solid echo pulse sequence (90°–τ–90°). The rate of decrease in the intensity of the ^{1}H NMR signal is determined by the strength of the magnetic couplings and the extent to which protons are static on a timescale of microseconds [SAK 93]. Protons separated by less than ~1 nm in rigid organic materials have strong static coupling. Their ^{1}H NMR signals exhibit short-lived, Gaussian-like distribution functions (relaxation time, T_2, of 16–24 μs) [SAK 93]. On the other hand, protons in mobile (liquid-like) molecular structures exhibit weaker static magnetic interactions. Their ^{1}H NMR signals decrease at a much slower rate and give long-lived protons, Lorentzian-like distribution functions (T_2 of 100–1,000 μs) [SAK 93].

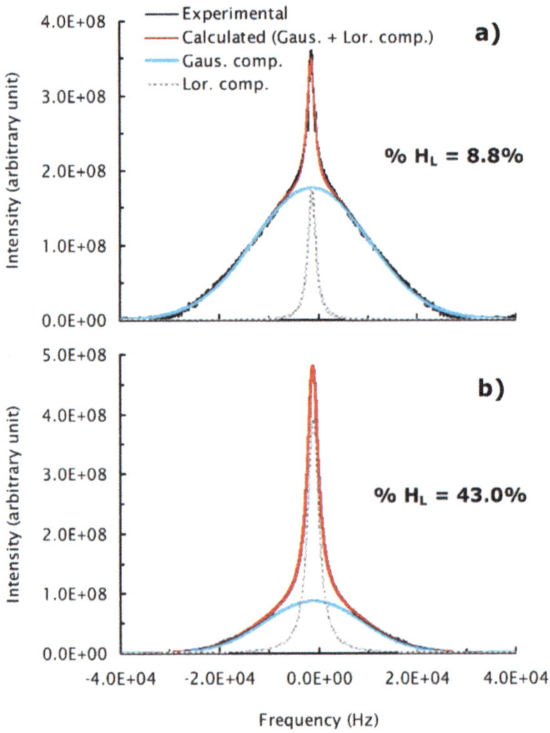

Figure 3.10. *^{1}H NMR spectra with deconvolutions into Gaussian and Lorentzian distributions and the calculated fraction of mobile hydrogen (%H_L) for Miscanthus at (a) 200°C and (b) 300°C heated at 5 K/min in the ^{1}H NMR probe [DUF 12a]. When mobile protons increase from 200°C to 300°C, the signal becomes "thinner" with a higher contribution in Lorentzian function. Calculated (red) and experimental (black) curves are on top of each other. For a color version of the figure, see www.iste.co.uk/dufour/biomass.zip*

The Fourier transform time-domain decays of the solid echo pulse signal were deconvoluted into Gaussian and Lorentzian distribution functions (see Figure 3.10) by a Matlab program developed by our group.

The program enables the calculation of the fraction of mobile hydrogen or "fluid phase" (%H_L) from the following relation:

$$\% \mathrm{H_L} = \frac{\mathrm{A_L}}{\mathrm{A_L} + \mathrm{A_G}}$$

where A_L and A_G are the areas of the Lorentzian and Gaussian functions, respectively.

When mobile protons increase as a result of an increase in temperature, as for *Miscanthus* between 200°C and 300°C in Figure 3.10, the NMR peak becomes "thinner" with a higher contribution from the Lorentzian component.

The evolution of fluid-phase development during pyrolysis of cellulose, xylan, lignin and *Miscanthus* is presented in Figure 3.11.

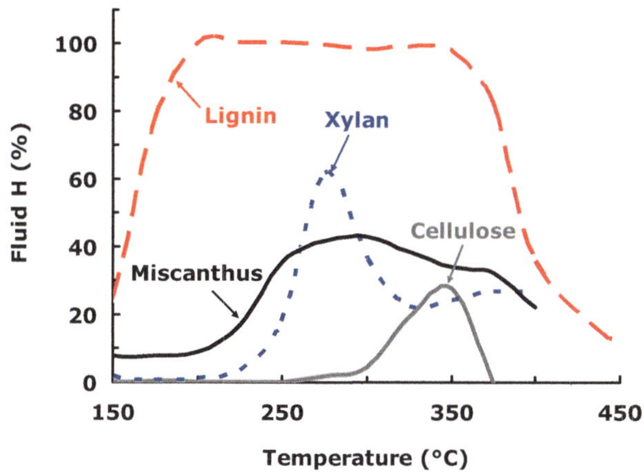

Figure 3.11. *Fluid-phase (%H_L mobile protons) development as a function of temperature. Sample heated in the ^{1}H NMR probe (5 K/min)*

The amount of fluid phase (%H_L) that the materials develop at their temperature of maximum fluidity decreases in the order: lignin (100% H_L) > xylan (61%) > *Miscanthus* (42%) > cellulose (28%).

The mobility development is shortly discussed for each individual material.

Lignin becomes completely fluid (100% H_L) at 200–225°C and remains completely fluid up to 350°C (Figure 3.11). At temperatures below 200°C, the glass transition of lignin is evidenced by the increase in fluidity without any mass loss and any thermal signal (from DSC analysis). The rubbery phase mobility is then overlapped by covalent bond scission. The chemical structure of lignin is affected by the thermal treatment above 200°C through homolytic cleavage of aryl ether linkages (mostly β-O-4 linkages) [NAK 08, BRO 10, FAR 10]. Small lignin fragments with a high molecular mobility are produced. With increasing temperature from 225°C to 350°C, the thermal depolymerization of lignin is associated with the well-known demethoxylation and recondensation reactions. At 350°C, the mobile fraction decreases due to the formation of solid compounds from crosslinking and growth of aromatic domains [WAN 90, NAK 08, FAR 10]. The decrease in the mobile H fraction evidenced at 350°C (Figure 3.11) is in accordance with the solidification stage within the lignin char observed at 350–400°C in [SHA 04].

Regarding cellulose, the proportion of mobile hydrogen is lower than that in lignin or xylan (Figure 3.11), but mobile protons from cellulose pyrolysis are evidenced for the first time even at low heating rates (5 K/min). The fluidity starts to increase slowly at 250°C, whereas mass loss starts at 290°C (see Figure 3.14) and increases sharply at 350°C. The observed mobile phase is probably not the "active cellulose" (cellulose with a low degree of polymerization of around 200 [BRO 73]). The mobile phase is probably developed by disordered carbohydrates [WOO 04] that can become liquid under pyrolysis conditions [MOK 83, LÉD 02, MAM 09].

Xylan develops a mobile phase, the importance of which increases from 200°C to 275°C (61% H_L, Figure 3.11). This is in agreement with microscopic observations indicating that xylan shows signs of melting or softening above 200°C [FIS 02]. Contrarily to lignin, the mobile phase from xylan is developed only after the beginning of mass loss (from 200°C). Consequently, the development of fluid material during xylan pyrolysis should not be due to a physical conversion of the native xylan but most probably from some pyrolysis products that are liquid at reaction temperature. There is still a lack of knowledge of xylan pyrolysis to discuss more in depth the origin of its fluidity.

Regarding the native polymer network in *Miscanthus*, its mobility starts at 200°C but before 150°C for the isolated lignin (Figure 3.11). The ^{1}H NMR results are in agreement with previous SEM observations, where lignin-rich xylem elements from tobacco did not "melt" until after 300°C whereas isolated lignin "molten" at a lower temperature showing interactions among components in the cell wall matrix [BAL 03]. Donohoe *et al.* [DON 08] showed that a dilute acid pretreatment causes lignins to coalesce into larger molten bodies that migrate within and out of the cell walls of maize and can redeposit on the surface of plant cell walls. Droplets of lignin have been evidenced at 150°C at the surface of maize cell walls. This temperature of lignin "softening" the native cell walls is lower than that in our case and that of [BAL 03], probably because of the aqueous dilute acid environment that promotes the mobility of lignin in the native network of the polymers.

It is important to note that the isolated lignin has a lower molecular mass than the native lignin [HAG 10] that could increase its mobility at a lower temperature.

The mobility starts always at a higher temperature and to a lower extent in native networks such as in *Miscanthus* than in synthetic blends. It reveals strong interactions between the polymers and their pyrolysis products inside the native network that are not evidenced by the additivity law on global mass loss (in accordance with our thermogravimetric study). These interactions could be explained from the structure of cell walls illustrated in Figure 3.12.

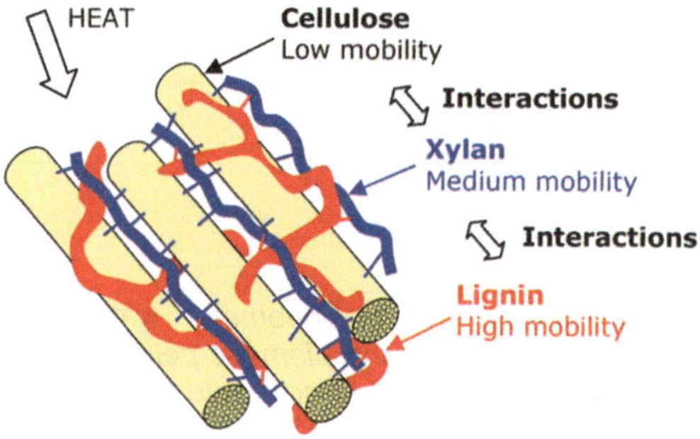

Figure 3.12. *Simplified scheme of the polymers network in the native biomass cell wall and mobility development during their thermal conversion [DUF 12b]. Cellulose "imposes" its rigidity to the polymer network during pyrolysis of biomass cell walls. For a color version of the figure, see www.iste.co.uk/dufour/biomass.zip*

The rigidity in cell walls arises from the crystallinity of the cellulose in the microfibril phase and from crosslinking of lignins in the matrix [NEW 92]. It could partly explain why the cellulose seems to "impose" its rigidity in the native network. When fractionating the biomass polymers, lignin and xylan give rise to higher mobility after extraction because they are no longer chemically linked with cellulose (even indirectly for lignin linked to xylan, xylan directly linked to cellulose). The beginning of mobility in *Miscanthus* (200°C) may be related to the beginning of xylan mass loss and/or to lignin–xylan bonds scission (ester and ether bonds) [DUF 12b].

3.3.2. *In situ analysis by rheology*

Figure 3.12 displays a simplified scheme of the method. Measurements were performed using a Rheometrics RDA-III high-torque controlled strain rheometer by our group (at CNRS-Nancy). The disk of biomass was placed between two serrated plates to reduce "slippage." Biomass between the plates and rheometer axe are inside a furnace flushed under N_2. Plates are heated through convection of N_2, and the temperature is monitored by a thermocouple placed inside the bottom plate. The furnace can be open allowing a "quench" of the sample at a specific temperature.

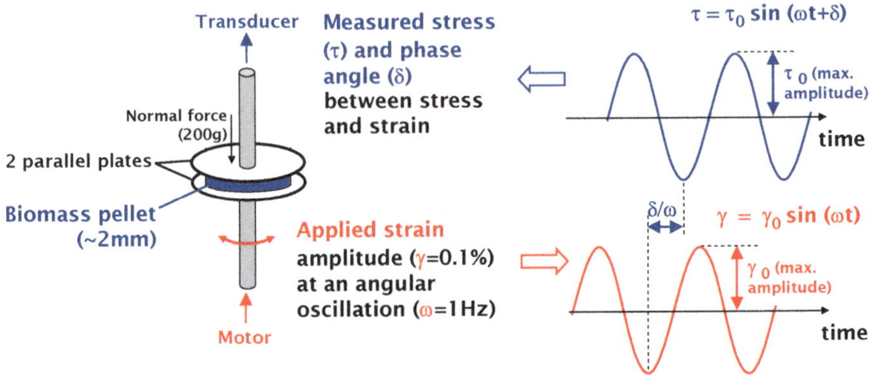

Figure 3.13. *Simplified scheme of the rheometer with imposed (in red) and measured (in blue) parameters. Plates and biomass pellet are flushed under N_2 and placed inside a furnace controlled by a thermocouple positioned in the bottom plate through the bottom axe. For a color version of the figure, see www.iste.co.uk/dufour/biomass.zip*

A very small amplitude angular oscillation was applied to the sample from the bottom plate (Figure 3.13). Tests were performed with 0.1% strain amplitude (γ, amplitude of oscillation divided by particle thickness) and a frequency (ω) of 1 Hz (6.28 rad/s) following the procedure optimized by [STE 04] to operate in the linear viscoelastic region of the material. The stress response (amplitude, τ_0, and phase angle, δ) was measured on the top plate to obtain G' and G''.

These parameters are calculated and defined as follows:

$$G' = \cos(\delta)\frac{\tau_0}{\gamma_0}$$

G' (Pa) is the elastic modulus. It is proportional to the mechanical elastic energy stored and reversibly recovered.

$$G'' = \sin(\delta)\frac{\tau_0}{\gamma_0}$$

G'' (Pa) is the viscous modulus. It is proportional to the mechanical energy irreversibly lost through viscous dissipation.

Throughout the test, a constant normal force of 200 g was applied to the sample from the top plate to reduce slippage. Therefore, the thickness of the sample was allowed to change. The thickness (ΔL) could be measured throughout the test and gives an indication of how much the sample is expanding/contracting (results not presented in this book, see [DUF 12a] for more details).

Elastic and viscous moduli for lignin (displayed between 50°C and 400°C), cellulose (150–400°C), xylan and *Miscanthus* are depicted with mass losses in Figure 3.14. All the materials were heated from room temperature to 400°C at 5 K/min in the rheometer, and the moduli were measured during the pyrolysis process. Mass losses were obtained from separate experiments in thermogravimetric analysis (TGA).

The glass transition phase of lignin is observed by rheology in agreement with the mobility development analyzed from ^{1}H NMR. Lignin becomes mainly viscous (viscous modulus higher than elastic modulus) at 130°C before any mass loss (Figure 3.14(a)). At about 140°C, both moduli decrease showing the formation of a soft material. Softening of lignin is due to the glass transition phase followed by bond breaking giving lower molecular weight species. The elastic modulus reaches its minimum value (~10^3 Pa) at 180°C whereas the viscous modulus at 220°C. Before 200°C, lignin is already degraded and darkened, in agreement with small mass loss. It is a soft and mainly viscous material.

At 200°C, elastic modulus becomes higher than viscous modulus (Figure 3.14(a)). The material is still soft, with low moduli, but becomes mainly elastic. Both moduli increase at 220°C. At 350°C, a fast solidification of the material occurs in close agreement with the decrease in mobile protons fractions evidenced by ^{1}H NMR (Figure 3.11). Crosslinking reactions lead to the formation of a solid matrix, mainly elastic and hard: a solid "char." At 380°C, the viscous modulus decreases, and the material becomes very elastic. A fluffy char is produced by volatiles that push the viscoelastic material and crosslinking reactions then harden the structure into a mainly elastic and less deformable material. The rheological behavior, combined with mass loss, is a very significant "signature" of thermal degradation of materials. We noticed different rheological signatures for different lignins that would depend on initial lignin structure (crosslinking, molecular weight, functional groups and heteroatoms).

Regarding cellulose, the evolutions of the viscous and elastic moduli are very different from those for lignin (Figure 3.14(b)). The material stays roughly hard (high moduli) through the degradation but both moduli decrease sharply at about 320°C before the temperature of maximum mass loss. A mild softening is thus evidenced even for this slow heating rate (5 K/min) consistent with the formation of mobile protons. Cellulose would even form a material with high viscous contribution during a short temperature range close to maximum mass loss temperature (335°C). Undoubtedly, a viscoelastic liquid could be formed at higher heating rates [LÉD 02, MAM 09].

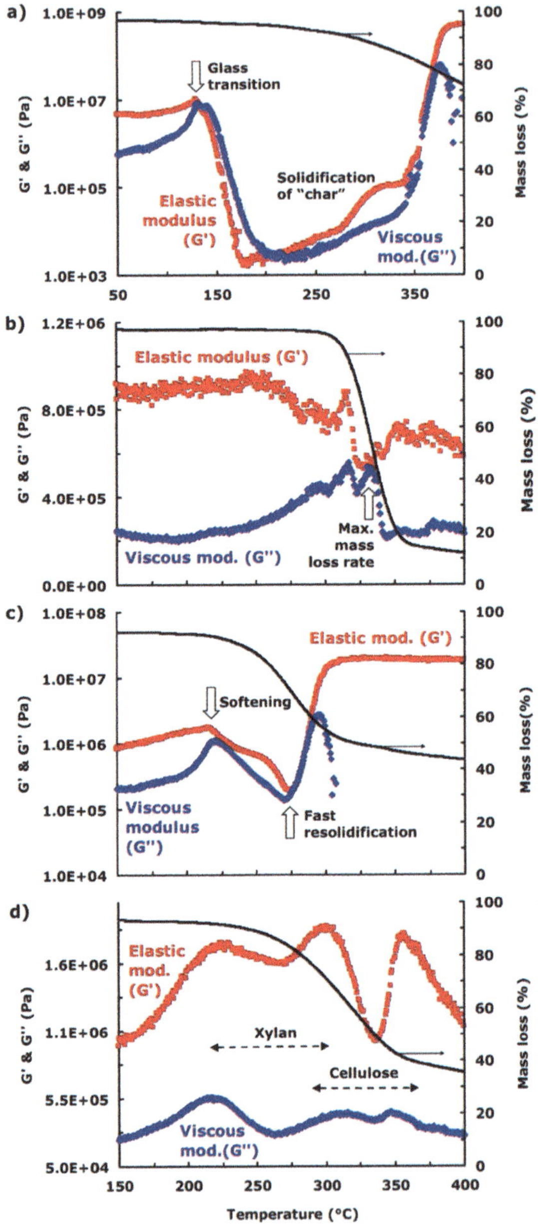

Figure 3.14. *Evolution of elastic (G') and viscous (G'') modulus (from high-temperature rheology) and mass loss (obtained by TGA) as a function of temperature (5 K/min) for (a) lignin (50–400°C), (b) cellulose (150–400°C), (c) xylan and (d) Miscanthus [DUF 12a]. For a color version of the figure, see www.iste.co.uk/dufour/biomass.zip*

Regarding xylan, softening starts at 220°C after the beginning of mass loss and solidification at 270°C at about 75% mass loss (on anhydrous basis, Figure 3.14(c)). It can be noticed that solidification is very fast just as for lignin and lead to a very elastic and hard char less fluffy and much harder than the lignin char.

Miscanthus (Figure 3.14(d)) exhibits very different viscoelastic properties than the other three materials. The contribution of xylan and cellulose can be seen at 270°C and 330°C, respectively, but lignin softening is not seen.

The elastic moduli of the four materials are compared in Figure 3.15.

The elastic modulus of the extracted lignin is three orders of magnitude lower than that of *Miscanthus* during softening and three orders of magnitude higher after solidification. However, the contribution of lignin to the rheological signature of *Miscanthus* is not captured. This finding is again in agreement with ^{1}H NMR analysis and with [BAL 03]. In the native network of biomass, lignin is linked to xylan and xylan to cellulose [JAR 00]. The rigidity of lignin fragments could be maintained by xylan before its thermal degradation and indirectly by cellulose. The elastic moduli of *Miscanthus* and cellulose have similar patterns on the logarithmic scale used in Figure 3.15. Cellulose imposes its viscoelastic behavior to the overall degraded polymer network from the native *Miscanthus*.

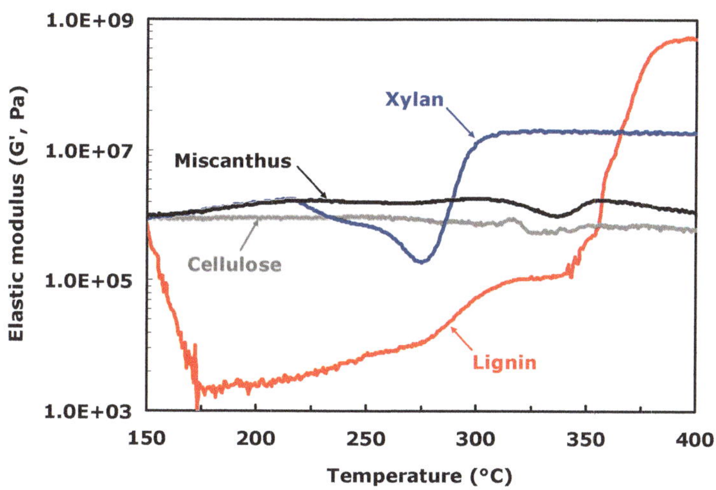

Figure 3.15. *Comparison of elastic modulus evolution as a function of temperature for lignin, cellulose, xylan and Miscanthus. The variation of elastic modulus for lignin is three orders of magnitude higher than that for Miscanthus. Nevertheless, lignin has no apparent contribution on the viscoelastic behavior of Miscanthus. For a color version of the figure, see www.iste.co.uk/dufour/biomass.zip*

The main findings from *in situ* rheology analysis are summed up in Figure 3.16.

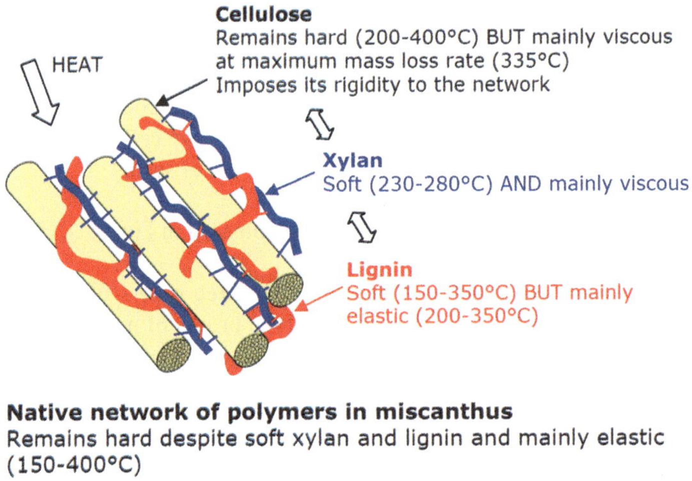

Figure 3.16. *Simplified scheme of polymer network in native biomass and summary of main findings from in-situ rheology analysis for the fractionated polymers and the native biomass [DUF 12a]. For a color version of the figure, see www.iste.co.uk/dufour/biomass.zip*

3.3.3. *Relations between ^{1}H NMR and rheology analysis*

Figure 3.17 displays the viscous modulus as a function of fluid-phase fraction. The physical analysis from rheology is thus compared with the chemical analysis from ^{1}H NMR.

It has been previously shown that the rheological properties of coal and some blends could be related with the mobile protons [DÍA 08]. Biomass and its constituents have a completely different pattern than coal because of the different origin in mobility formation between biomass constituents and coal.

Figure 3.17(a) shows an almost linear relationship during softening of xylan and lignin but, when solidification proceeds, mobile protons remain in the elastic solid-like char. This finding is in agreement with that of Haas *et al.* [HAA 09] who showed that liquid pyrolysis products are entrapped in heat-treated wood. To the best of our knowledge, we provide the first real-time high-temperature quantification of formed and entrapped mobile products.

It is evidenced that molecular mobility is formed during cellulose pyrolysis in a mainly elastic material with few variations in elastic modulus compared with mobile protons (Figure 3.17(b)). This finding is in agreement and supports the concept of "cavities" extensively developed by Mamleev *et al.* [MAM 09]. High-boiling liquid tar is encapsulated in the cavities (nuclei) and launches the ionic mechanisms occurring in cellulose pyrolysis [MAM 09]. Furthermore, the materials can present dispersions or suspensions. The liquid phase can fill up an interfibrillar space and coat solid particles [DON 08].

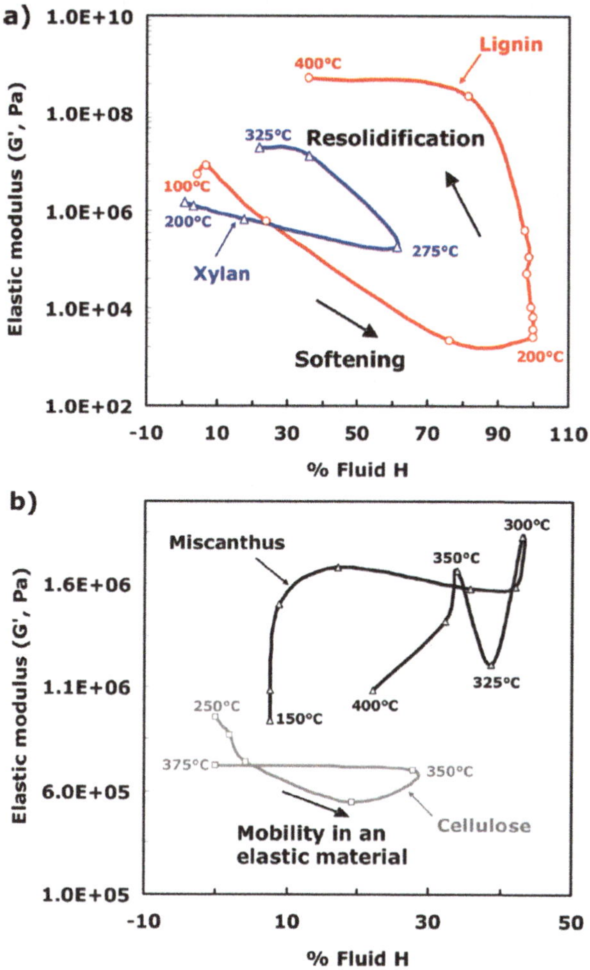

Figure 3.17. *Evolution of elastic modulus (G′, from rheology analysis) as a function of mobile protons (% fluid H, from ^1H NMR analysis) for (a) xylan and lignin and (b) cellulose and Miscanthus. Mobility (% fluid H) is still present after solidification of lignin and xylan. Mobility of cellulose occurs in an elastic material, in solid-like cavities*

Regarding *Miscanthus*, it can be seen from Figure 3.17(b) that native biomass leads to a much more complex behavior than separated polymers. Further investigations are needed with other complementary methods such as ^{13}C NMR to better understand the intermediate materials formed in the network of polymers/char.

3.4. Structure of a char particle and mechanisms of its oxidation

Pyrolysis of biomass yields a complex viscoelastic structure (from rheology study), which in turn yields a hard material: the char. Depending on pyrolysis conditions and biomass type (especially mineral matters), the char can exhibit different porous structures and reactivity toward oxidation (by O_2, H_2O or CO_2). The char globally keeps the macroporous structure of wood as evidenced by SEM (Figure 3.3). But, at the molecular level, it consists of disordered aromatic sheets resembling a "crumpled paper" [STO 90] formed by crosslinking reactions in cell walls (Figure 3.18).

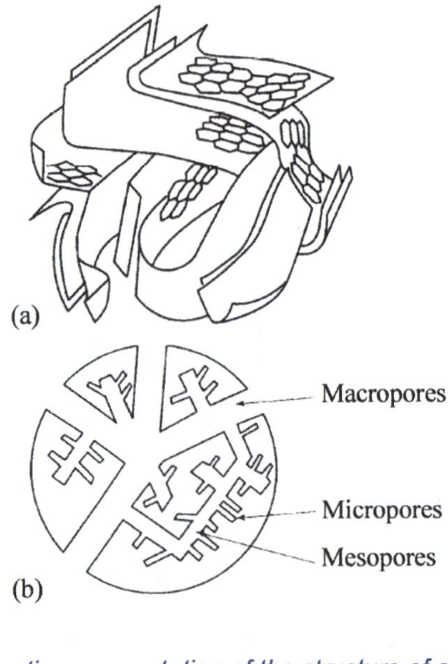

Figure 3.18. *Schematic representation of the structure of char: (a) disorganized nanostructure (graphene sheets) and (b) porous structure [STO 90]*

The structure of a char consists of aromatic sheets and strips, often bent and resembling a mixture of crumpled paper, with variable gaps of molecular dimensions between them. These gaps are the micropores. A wood char is highly disorganized.

The porosity of the char can be defined by three classes: (1) the macropores are defined for pores larger than 50 nm diameter, (2) mesopores from 2 to 50 nm and (3) micropores defined for pores up to 2 nm.

The nanostructure of char can be well analyzed by high-resolution transmission electron miscroscopy (TEM) [ROU 90] and is presented in Figure 3.19.

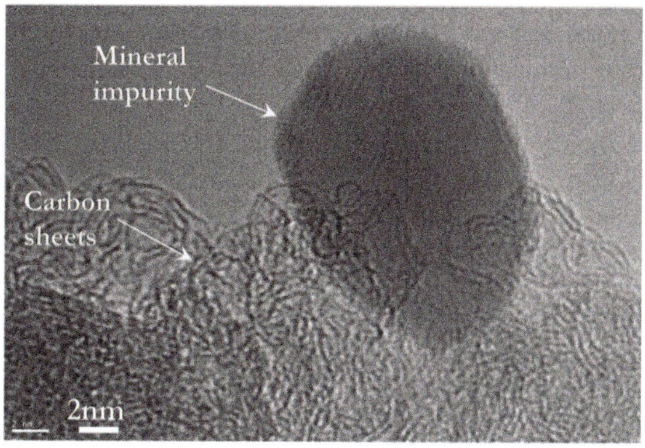

Figure 3.19. *Nanostructure of wood char analyzed high-resolution TEM [DUF 15]*

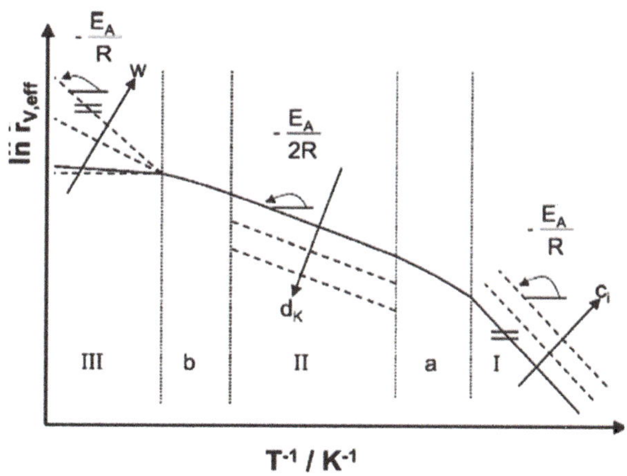

Figure 3.20. *The three regimes for gas/solid reactions applied to char oxidation (a specific gas/solid reaction)*

The literature proposes a huge numbers of kinetic constant to model the reactions of char oxidation (C + O_2 or +H_2O or +CO_2) [RAD 91, KLO 05]. The important variation in the kinetic constants may be explained by the various compositions or structure of chars. Indeed, minerals present first as nutrients in the wood and then in various forms in the char [DUF 15] catalyze the reactions of char oxidation. Another reason may be that the experiments used to derive the kinetic constants are not conducted in the chemical regime. It is well known that all gas/solid reactions can process following three regimes presented in Figure 3.20.

In zone I of Figure 3.20 (at low temperature), the apparent rate of char oxidation is controlled by the chemical kinetic of the reaction. It is the "chemical regime."

In zone II (at intermediate temperature), the apparent rate of char oxidation is controlled by gas (e.g. O_2 or H_2O) diffusion (d_K, Knudsen diffusion regime) inside the pores of the char particles. Thus, the apparent rate of char consumption depends on the particle size and/or its porous structure.

In zone III, the apparent conversion rate of the particle is limited by the external mass transfer of the surrounding reactive gas (e.g. O_2 or H_2O) to the outer surface of the particle. The apparent rate of char oxidation depends on the velocity of the surrounding gas in the bed of particles.

Numerous detailed models for describing char oxidation within these three regimes were developed in the literature [HAU 14].

When the regime is chemically controlled, that is without mass transfer limitations (for small particles, low temperature and high reactive gas velocity), the reactivity of a char toward oxidation is correlated with its active surface area (ASA), corresponding to the surface area occupied by the oxygen chemisorbed at 300°C and 66 Pa initial oxygen pressure [LAI 63, RAD 83, EHR 89]. The active sites are carbon atoms with dangling bonds, located at the edges of the carbon layers, at defect points such as vacancies, dislocations and steps on the outer basal plane [RAD 83].

A char oxidation model was developed by our group [ZOU 15], which handles for the first time the evolution of ASA and mass transfers as a function of char particles conversion. This model is not presented here for sake of brevity. As the oxidation of char proceeds, pores are opened and the increase in the surface area can be mainly attributed to the opening of the microporosity (<2 nm). Knudsen diffusion in micropores can be the limiting mechanism during char oxidation. The model also handles extraparticle mass transfer of the oxidant gas by convection. The model has been validated based on the literature data [KLO 05].

4

Molecular Scale

4.1. Introduction on the variability of biomass composition and the decoupling of reactions

In Chapter 3, the main intraparticle mechanisms occurring during biomass pyrolysis were presented. We have explained how an intermediate liquid phase is formed during pyrolysis to produce "gas, tar and char." But at the molecular level, the mechanism is much more complex (Figure 4.1). The "intermediate liquid" is formed by a complex network of polymers and minerals, which yields a pool of intermediate compounds and generates hundreds of volatile species (primary tars and permanent gases).

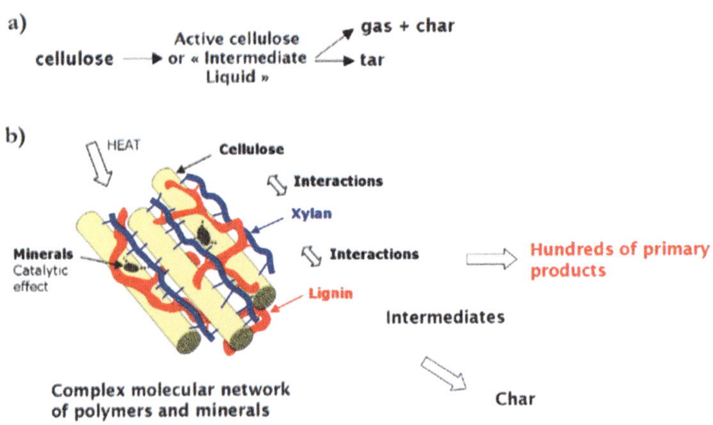

Figure 4.1. *Biomass primary pyrolysis: from (a) simplified mechanism to (b) complex molecular mechanism. For a color version of the figure, see www.iste.co.uk/dufour/biomass.zip*

The composition of pyrolysis products highly depends on the composition of biomass, especially on minerals that have an important catalytic effect.

In this chapter, first we introduce the molecular structure of biomass.

A decoupling methodology is required to better understand the chemical mechanisms occurring in different reactors. For instance, if the temperature in a fluidized bed is modified, the primary pyrolysis and the secondary reactions (volatiles conversion) would be both affected.

Therefore, we describe first the chemical mechanisms of primary pyrolysis based on an exhaustive review of the literature and experimental work conducted by our group. These complex mechanisms are shown by various analytical methods.

Then the secondary reactions of primary tar conversion are explained.

4.2. Chemical structure of lignocellulosic biomass

Lignocellulosic biomass (including woody and herbaceous biomass) is composed of three macromolecules: cellulose (40–60%mas.), hemicelluloses (10–30%mas.) and lignin (20–25%), which are organized in a complex network in the cell wall [ZAK 10].

Cellulose is a polymer of β-D-glucopyranose. Its structure is presented in Figure 4.2. It is a polymer based on the cellobiose unit and with a high degree of polymerization. Cellobiose comprises two β-D-glucopyranose linked by β(1,4) bonds. The formula of cellulose is $(C_6H_{10}O_5)^n$. It is the most abundant polymer on Earth (43–60%mas. of vegetal biomass).

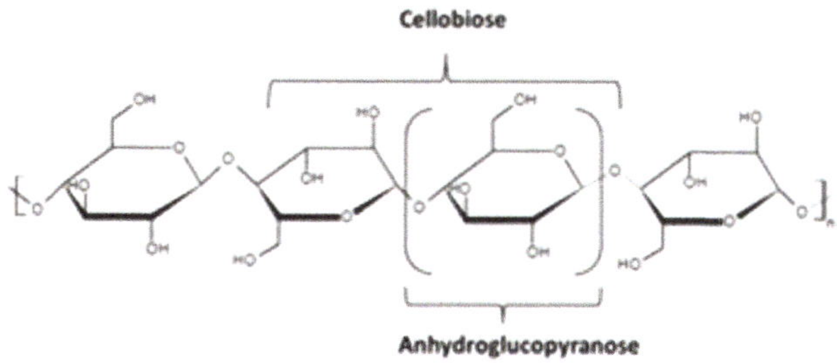

Figure 4.2. *Chemical structure of cellulose*

The hemicelluloses are carbohydrate macromolecules with a lower degree of polymerization than that of cellulose. Their composition depends on the type of biomass and are a mixture of various monosaccharides linked by osidic bonds (β(1,4), β(1,3), α(1,2), etc.). The chemical structures of the monosaccharides are given in Figure 4.3.

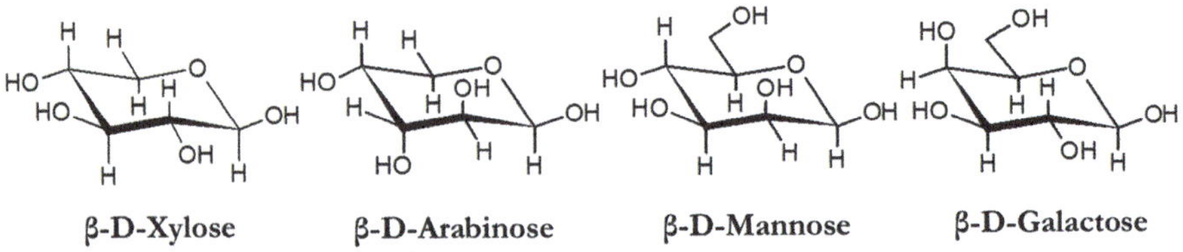

Figure 4.3. *Chemical structure of hemicelluloses*

The composition in monosaccharides of various biomasses is reported in Table 4.1.

	Glucose[a]	Xylose	Mannose	Galactose	Arabinose
Wheat straw	42,7	23.1	0	–	–
Miscanthus	48,5	18.1	0	–	–
Red oak	41	19	2	1.2	0.4
Norway spruce	43	7.4	9.5	2.3	1.4
Douglas fir	44	2.8	11	4.7	2.7

[a]Including glucose in cellulose.

Table 4.1. *Composition (%dry mass) in sugars of various biomasses [LEB 15a]*

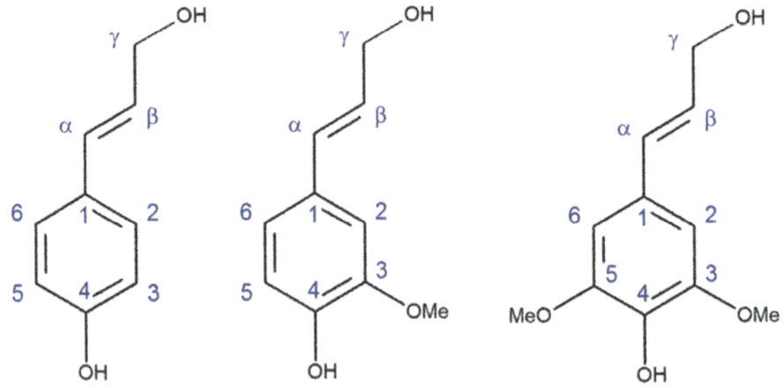

Figure 4.4. *The three monolignols, the subunits of lignin*

The third macromolecule in cell walls is lignin. Its chemical structure comprises phenolic unit but it is not yet well defined. Lignin is composed of three phenolic subunits presented in Figure 4.4.

These subunits are linked by various bonds: ether (β-O-4, α-O-4 and 4-O-5), carbon–carbon (β-5, β-β, 5-5′ and β-1) (Figure 4.5) [LEB 15a].

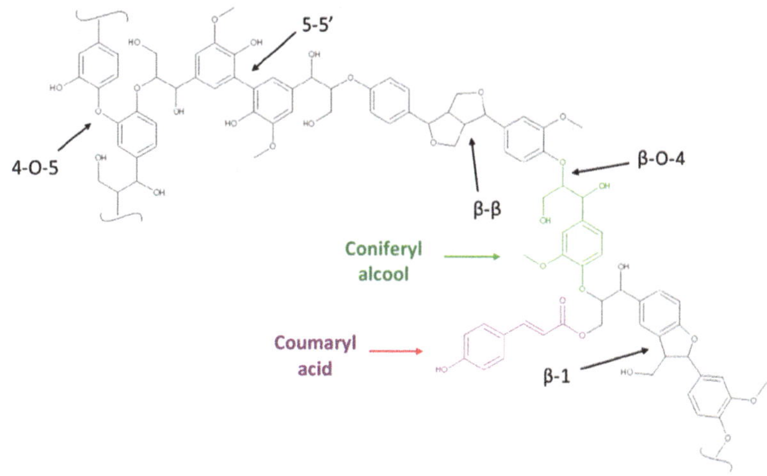

Figure 4.5. *An example of a part of the structure of lignin from softwood (as fir or spruce) [LEB 15a]. For a color version of the figure, see www.iste.co.uk/dufour/biomass.zip*

4.3. Chemical mechanisms of primary pyrolysis

4.3.1. *Chemical mechanism of cellulose pyrolysis*

The mechanisms of cellulose pyrolysis proposed in the literature have been comprehensively reviewed by our group [LEB 15a]. Mechanisms are proposed since the early 20th Century. The primary conversion of cellulose occurs principally from 300°C to 400°C following different chemical mechanisms that are summarized in Figure 4.6. These mechanisms are favored or inhibited by the operating conditions (temperature, heat flux density) or the presence of inorganics (especially potassium, Figure 4.7). Inorganics (or nutrients, minerals, present naturally in biomass) have an important impact on pyrolysis product distribution due to their catalytic effects. Cellulose can be converted by three main pathways:

– transglycosylation [PIS 86, EVA 87, MAM 09], which leads to the formation of anhydro-sugars (levoglucosan);

– dehydration [KIL 65, SCH 01, CHA 09], which induces the formation of unsaturated bonds;

– open-ring/fragmentation [RIC 87, PIS 89, SHE 09] reactions, which lead to the formation of light oxygenated compounds.

The compounds formed from cellulose (furans, methylglyoxal, etc.) are produced by the combination of these different mechanisms. For instance the formation of furans needs open-ring reactions and dehydration in order to create furan ring from the glucopyranose structure.

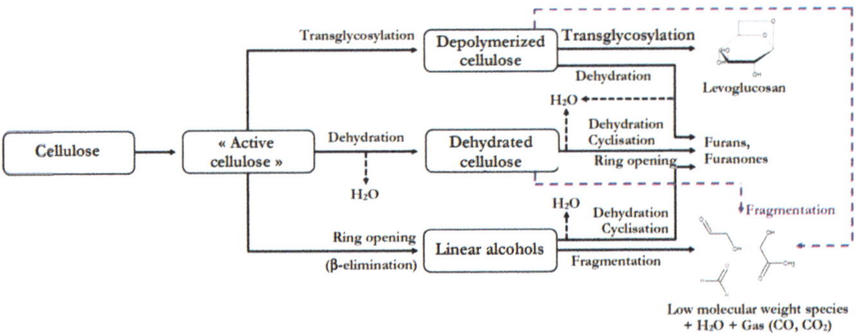

Figure 4.6. *Global mechanism for cellulose pyrolysis proposed by our group based on a comprehensive literature review (Le Brech, 2015)*

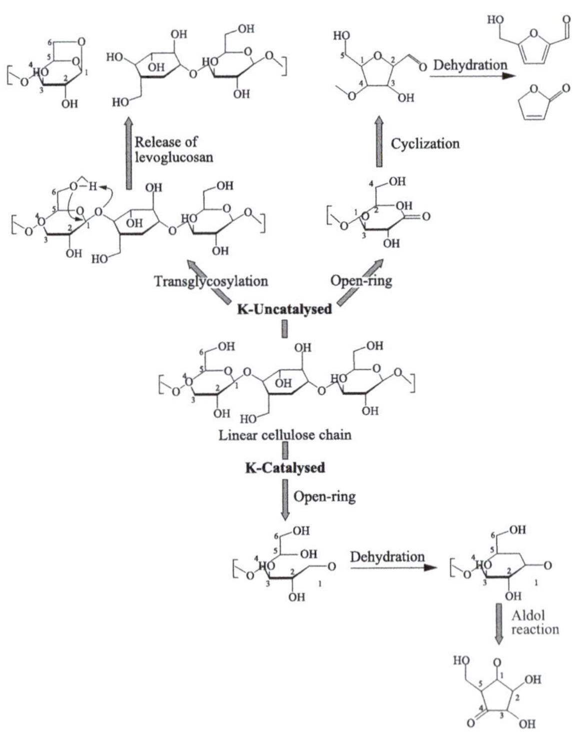

Figure 4.7. *Effect of potassium on the chemical mechanism of cellulose pyrolysis*

4.3.2. *Chemical mechanism of hemicellulose pyrolysis*

Degradation of hemicelluloses occurs on a broader range of temperature than that of cellulose (from 200°C to 350°C) because hemicelluloses are formed by different molecules linked by various chemical bonds that induce different thermal stabilities [WER 14]. Xylan (the polymer of xylose) is the most studied polymer for hemicellulose pyrolysis [WOR 07, SHE 10a]. At low temperature (200–250°C), hemicelluloses undergo dehydration reactions [COL 14] and specific chemical function fragmentations [SHE 10a]. Acetyl, carboxylic and methoxyl moieties are converted to formic acid, acetic acid and methanol, respectively [PEN 10, SHE 10a]. At higher temperature (250–350°C), the formation of furanic compounds (furans and furanones) is observed from the combination of dehydration, open-ring and cyclization reactions. The formation of anhydro-sugars (levoglucosan, levogalactosan or 1,4-anhydro-D-xylopyranose, etc.) from glycosidic ruptures is also described [HOS 07]. Furfural is one of the major compounds from xylan pyrolysis [SHE 10a]. Pyranones are also described as a product of hemicellulose pyrolysis [OHN 77, PON 91] (Figure 4.8).

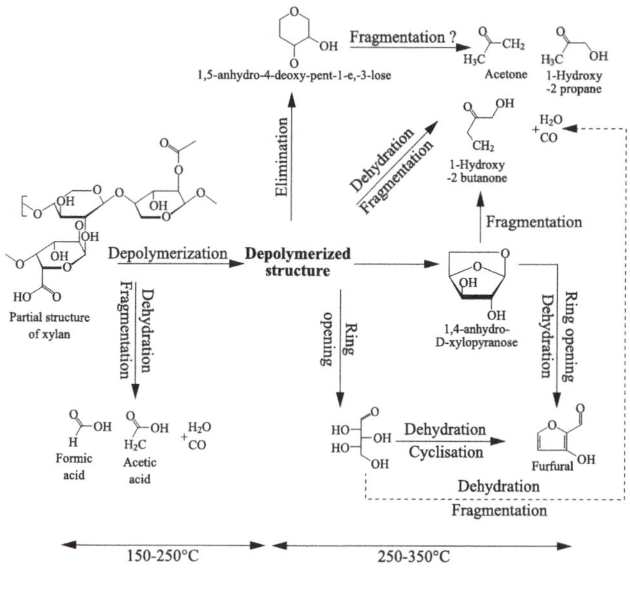

Figure 4.8. *Global mechanism for a partial structure of xylan (one constituent of hemicelluloses) pyrolysis proposed by our group based on a comprehensive literature review (Le Brech, 2015)*

4.3.3. *Chemical mechanism of lignin pyrolysis*

Lignin is a phenolic macromolecule composed of phenylpropane units linked together through various ether and carbon–carbon linkages [KAW 07, PAN 11]. The main conversion step of lignin occurs over a large temperature range from 150°C to 500°C [JAK 95, WAN 09]. At low temperature (150–300°C), oxygenated propyl chains are mainly converted to water by dehydration reactions [JAK 95] and formaldehydes by the release of carbonyl functions localized on carbon Cγ [KAW 07] (Figure 4.9).

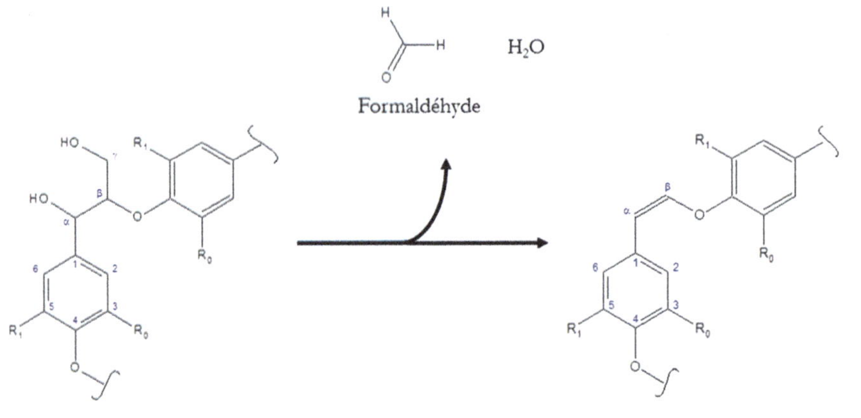

Figure 4.9. *Mechanism of lignin conversion by Cγ elimination and dehydration of propyl chains*

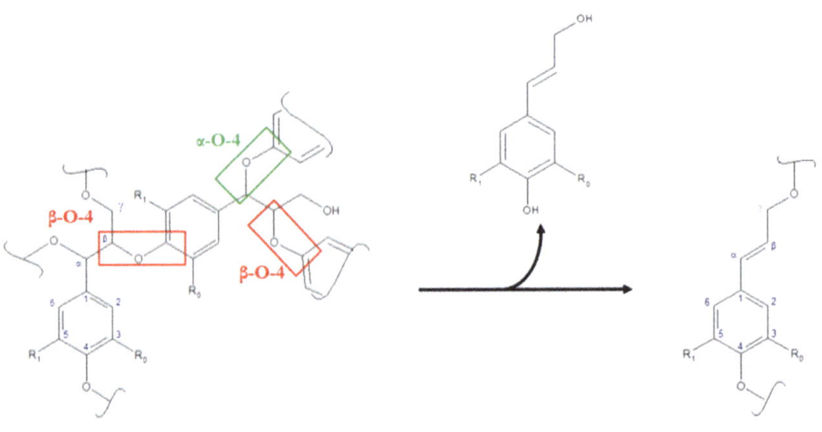

Figure 4.10. *Mechanism of ether linkages conversion during lignin pyrolysis. For a color version of the figure, see www.iste.co.uk/dufour/biomass.zip*

The ether linkages (α-O-4 and β-O-4) can react at 200°C [COL 14] leading to the formation of unsaturated chains or to the release of phenolic compounds with chemical structures close to the native network. Moreover, these phenolic compounds are composed of oxygenated aromatic rings substituted by an alkyl chain with two or three carbons (eugenol, 4-vinylguaiacol) (Figure 4.10).

At temperatures higher than 300°C, alkyl chain conversion is global [SHA 04] and phenolic compounds with or without a short alkyl chain (mainly one carbon such as methylguaiacol) are devolatilized [EVA 87, SHE 10a, DUF 13]. Methoxyl moieties become reactive at approximately 370°C (depending on the inorganic content of lignins), leading to the formation of methanol and methane by demethoxylation and demethylation reactions, respectively. Consequently, demethoxylated (phenol, methylphenol) and demethylated (catechol, cresol) phenolic compounds are also released (Figure 4.11).

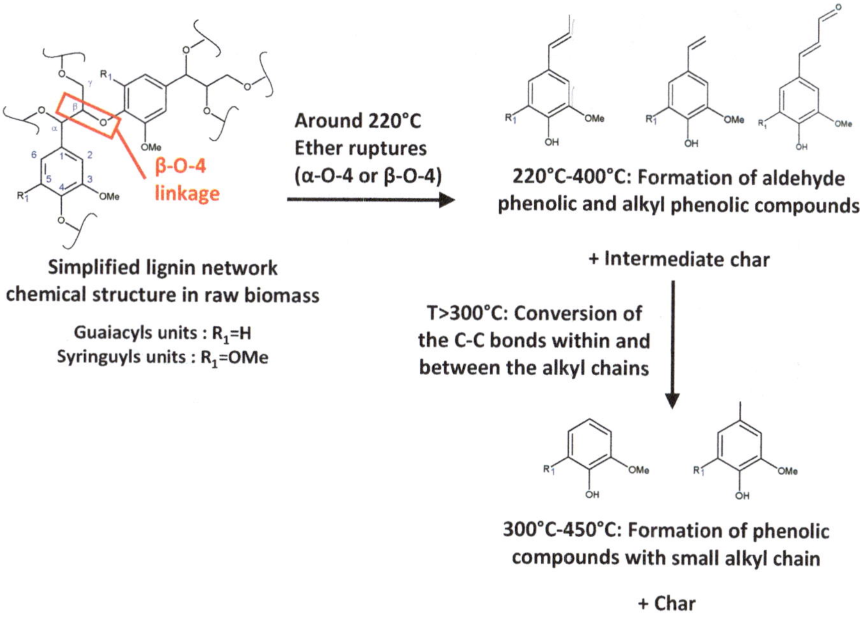

Figure 4.11. *Global simplified mechanism of lignin pyrolysis [LEB 15a]*

4.3.4. *Analysis of biomass primary tar by online soft photoionization*

The primary oxygenated species released by biomass primary pyrolysis are very fragile species and this is interesting to analyze them online (in real time and temperature) by a soft method that does not fragment these compounds.

Our group has developed an advanced online soft photoionization mass spectrometry (MS) analysis of primary tar in collaboration with Rostock University (Germany) and USTC (Hefei, China) [FEN 12, DUF 13].

Figure 4.12 gives an example of a result obtained by online analysis of tar by this method during the primary slow pyrolysis of *Miscanthus* (a type of biomass).

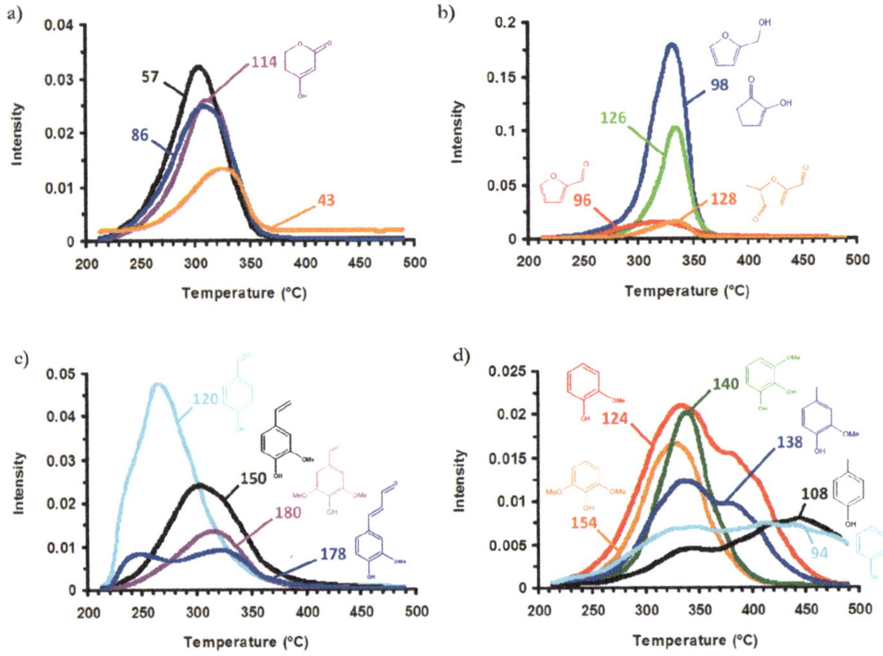

Figure 4.12. *Online analysis of key markers by soft photoionization MS as a function of temperature for Miscanthus pyrolysis: (a) hemicelluloses and carbohydrate fragments,(b) main cellulose markers, (c) lignin markers formed at low temperature and (d) lignin markers formed at high temperature [LEB 15a]. For a color version of the figure, see www.iste.co.uk/dufour/biomass.zip*

This method is a very powerful tool to reveal the temperatures of formation of the markers (primary tars) from cellulose, hemicellulose and lignin pyrolysis.

4.3.5. *Analysis of bio-oils by high-resolution mass spectrometry and GC*GC/MS-FID*

After condensation, primary tars yield "bio-oils" that are composed of hundreds of compounds. In our group, we have combined high-resolution MS, called "petroleomic" method, and gas chromatography/mass spectrometry (GC/MS) to analyze the majority of these compounds [OLC 13b].

The following two analytical methods are very complementary:

– a high-resolution Fourier transform ion cyclotron resonance (FT-ICR, 9.4 T) MS, which is used for heavier species (*m/z* of about 200 Da);

– quantitative GC*GC (heart cutting)/MS-flame ionization detector (FID) analysis, which is used for lighter species.

The methodology is illustrated on methanol-soluble bio-oils produced by lignin pyrolysis (Figure 4.13).

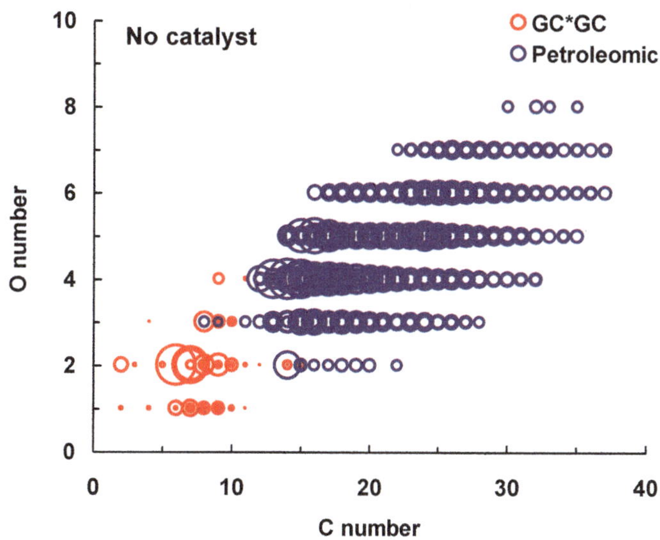

Figure 4.13. *Compound class (C and O numbers) determined by GC*GC and petroleomic analysis (LDI) as a function of weight yield (for GC) and relative abundance for petroleomic, for lignin pyrolysis products soluble in methanol. The area of circles is a function of weight yield for species analyzed by GC and relative abundance for petroleomic and cannot be compared between the two analytical methods. It can be compared only between the two experimental conditions and for the same analytical method. For a color version of the figure, see www.iste.co.uk/dufour/biomass.zip*

A much broader range of species is analyzed by petroleomics: from O_2 to O_8 and C_8 to C_{37} in lignin products (Figure 4.13) whereas species up to O_4 and C_{14} are analyzed by GC.

GC*GC analysis is calibrated by a combination of internal standard and prediction of response factors on the FID. A single injection allows quantifying about 100 compounds.

Laser desorption ionization (LDI) and electrospray ionization (ESI) were compared for the petroleomic analysis (not shown). A large number of peaks were detected, close to 300 peaks by ESI FT-ICR MS and more than 600 peaks by LDI FT-ICR MS. LDI efficiently ionizes a broader range of species than ESI. ESI and LDI have to be seen as complementary ionization techniques, which reveal different classes of components: highly polar components by ESI and less polar, more conjugated, heavier and on a broader range of *m/z* by LDI.

This approach shows that hundreds of molecules (especially heavier than 200 Da) are not commonly analyzed in bio-oils by conventional methods (GC/MS analysis).

4.3.6. *Analysis of biomass and biochar by solid-state nuclear magnetic resonance*

The analysis of tar (by on-line or off-line methods) is not sufficient to understand fully the mechanisms of biomass pyrolysis. One should also analyze the chemical composition of the solid residue (char).

Nuclear magnetic resonance (NMR) methods are non-destructive and have been used in many investigations for the analysis of coal, biomass polymers or char [HAW 84, TEK 87, HAT 88, LOV 93, KON 02, SIV 02, BAR 07, SAL 09, VIL 09, CAO 12, MAO 12, MEL 12, FOS 14]. Some NMR characterizations were performed using isolated fractions of the plant cell wall polymers, which require tedious separation and isolation techniques of cellulose, lignin and hemicellulose. This approach is not only time-consuming but also likely to alter the native structure of the polymers. Recently, the solution-state NMR spectroscopy of whole plant cell was reported by using ionic liquids [SAM 11, FOS 12], acetylated cell wall or swelled cell wall in a solvent to produce gel [KIM 10, WEN 13]. These methods require long preparation time and can alter the chemical structure of native polymers as well (e.g. lignin in ionic liquids). In addition, they are not adapted to study the chemical structure of cross-linked materials such as char that cannot be swelled by a solvent without chemical modification. Solid-state NMR is a promising method for the analysis of chemical structures of insoluble materials. However, one-dimensional (1D) solid-state NMR applied to whole plant material or to char is often limited by poor resolution and overlapping caused by the complex structure of the biomass. Advanced two-dimensional (2D) solid-state NMR methods have been previously used for biomass in extensive work [BAR 97, BAR 09, MAO 10] but at relatively low field.

For these reasons, we have used in our group a high-resolution solid-state 2D NMR method in collaboration with the universities of Mulhouse and Strasbourg (France). These methods have been applied to biomass and biochar for revealing the mechanism of biomass pyrolysis.

An ^{1}H-^{13}C 2D heteronuclear correlation (HETCOR) experiment with frequency-switched Lee–Goldburg (FSLG) irradiation has been performed using a high-field spectrometer (750 MHz). This method leads to previously unattained resolution for biomass and biochar and offers a unique ability to reveal their chemical composition. The dramatic increase in resolution in the proton dimension obtained by FSLG homonuclear decoupling together with a high field makes this technique a unique choice for proton-carbon analysis in the solid state [LEB 15b].

The different kinds of sugar and lignin carbons analyzed by this method are named in a standard way (Figure 4.14). The whole cell wall of *Miscanthus* straw was analyzed in situ by solid-state FSLG ^{1}H-^{13}C HETCOR (Figure 4.15).

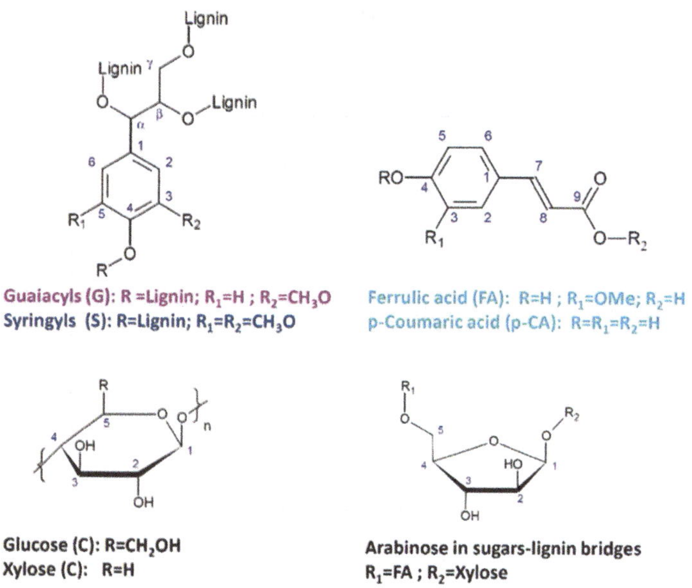

Figure 4.14. *Schematic representation of the main moieties of Miscanthus giganteus and notations for carbon atoms*

The HETCOR NMR spectrum of untreated *Miscanthus* yielded characteristic cross peaks for all components, namely, cellulose, hemicellulose and lignin. The clusters of cross peaks at δC/δH 64.4/3.6, 74.9–72.3/3.6, 83.6/3.6, 87.9/3.6 and 104.8/3.6 are assigned to C6 (CH2OH in glucose unit), C2–3, C5, C4 disordered structure, C4 ordered structure and C1 carbons of polysaccharides, respectively. These carbons seem to correlate with their own protons [MAO 10]. But in this range of 13C chemical shift (105–64 ppm), there is also a 13C signal from aliphatic lignin [LII 00]. Correlation between these carbons with O-alkyl protons and aromatic protons is shown in the area of δC/δH 50.0–110/3–7 ppm. Concerning lignin, correlations are observed at δC/δH 56.2/3.6 and 56.2/6.35–7 ppm corresponding, respectively, to O-alkyl protons from CH_3O and aromatic protons from lignin aromatic rings.

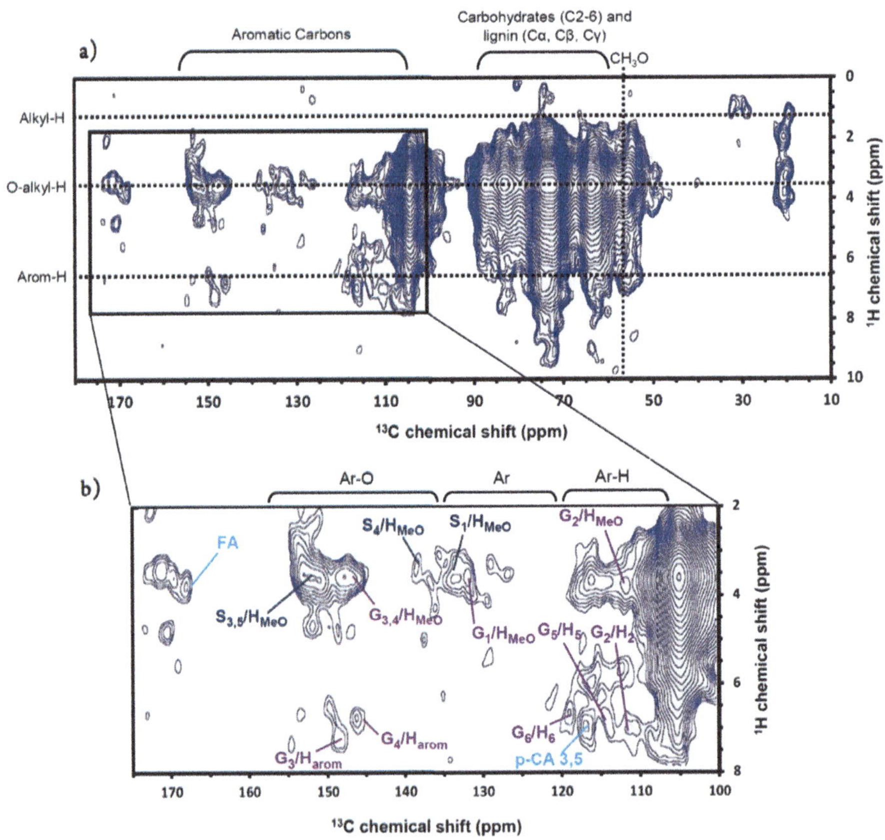

Figure 4.15. *(a) ^{13}C-^{1}H 2D HETCOR spectra of Miscanthus, (b) zoom on 175–100 ppm zone. For a color version of the figure, see www.iste.co.uk/dufour/biomass.zip*

According to the ^{13}C chemical shift 74 and 104 ppm, correlation with aromatic protons (6–7 ppm) seems due to interaction between carbohydrate carbons and lignin aromatic protons [MAU 02]. Indeed solid-state HETCOR measurement allows characterizing correlations between carbon and proton through space. These correlations indicate that carbohydrates are closely associated with lignin in biomass structures.

A *para*-coumarate group was also detected at δC/δH 116.1/6.9 ppm (*p*-CA3 and *p*-CA5). Signals for correlations of S1 and G1 units (aromatic carbons) with aliphatic CH–O protons were detected at δC/δH 132–136/3.0–4.0 ppm. Cross signals of aromatic C–O carbons with aliphatic CH–O protons corresponding to S4, G3-4 and S3-5 were observed at δC/δH 138.0–140.0/3.2–3.8, 146.0–149.5/3.1–4.0 and 150.0–154.0/3.2–4.0 ppm, respectively. Long-range correlations between aromatic C–O carbons with aromatic protons were also observed at δC/δH 145–147/6.8–7.2 ppm (G4) and δC/δH 147–149/7.0–7.5 ppm (G3).

The spectra of native *Miscanthus* and char produced at 300°C are compared in Figure 4.16. The pyrolysis of *Miscanthus* was conducted in a vertical U-shaped quartz fixed-bed reactor as those used in catalytic studies. The *Miscanthus* particles (800 mg, sieved between 40 and 100 µm) were supported on a quartz sintered plate (20 mm internal diameter). This geometry allows a good control of temperature, with a thermocouple inside the fixed bed of biomass and of mass transfer by flushing the fixed bed with a given flow rate of Argon (300 NmL/min). The reactor was heated by an electrical furnace from 20°C to 300°C at a heating rate of 5 K/min. Once 300°C was reached, the furnace was immediately moved down and the reactor was immediately immersed in water. The cooling time from 300°C to 200°C (thermocouple inside the fixed bed) was about 30 s. Char was cooled under Argon down to 30°C and then collected for further analysis. The char yield was 63%mas. This reactor is commonly used in our group to study the effect of biomass types and pyrolysis final temperature on mass balance (gas, condensable, char) and compositions of tar and char (by NMR).

The cross-linked network produced in char requires the solid-state NMR to understand the chemical mechanisms involved in thermal conversion. The high resolution of our NMR method enables us to reveal some typical chemical moieties in the char (Figure 4.17), in agreement with previous work [BAC 09, CHU 13].

The conversion of carbonyl functions is observed in the region of 165–173 ppm, indicating the cleavage between ferulic acid (in lignin) and arabinose (Figure 4.14) in the network. The shoulder at around 100 ppm on the C1 sugars strongly decreases pointing out the degradation of hemicelluloses (mainly xylan) in agreement with previous work [INA 07, MEL 12] (Figure 4.16(a)).

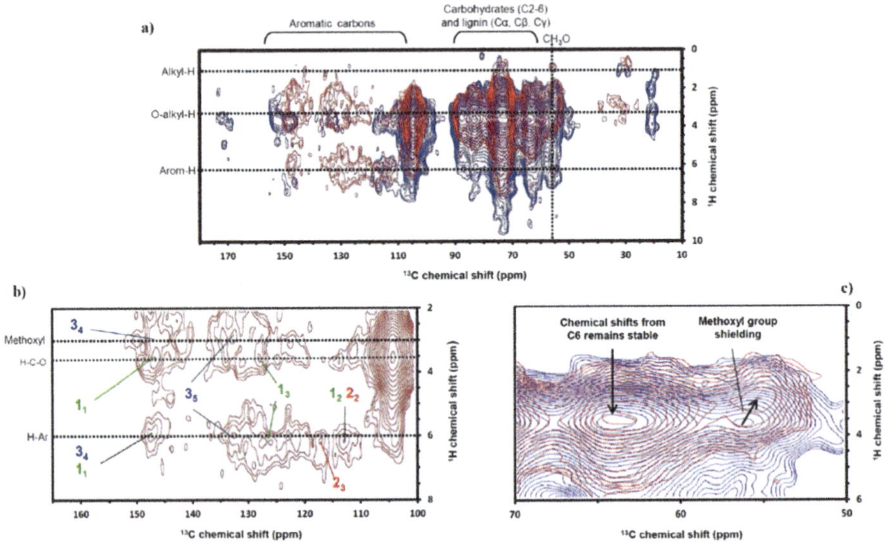

Figure 4.16. *^{13}C-^1H 2D HETCOR spectra: (a) comparison between native Miscanthus and biochar produced at 300°C, (b) zoom on 165–100 ppm $\delta^{13}C$ zone of biochar and (c) zoom on 70–50 ppm $\delta^{13}C$ comparison between native Miscanthus and biochar highlighting the shielding of methoxyl group in lignin-based aromatic clusters. For a color version of the figure, see www.iste.co.uk/dufour/biomass.zip*

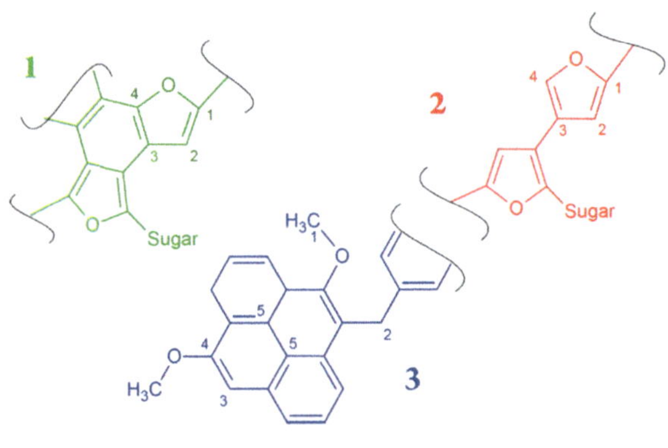

Figure 4.17. *Schematic representation of some chemical moieties in char*

Regarding the aromatic carbons, signals from S3 and S5 are no more present at 300°C. This can be explained by the cleavage of β-O-4 functions or by demethoxylation of S5 carbons [MEL 12]. The chemical shifts of the cross signals 167–168/3–4.0 ppm (Figure 4.16(a)) could be assigned to methoxylated aromatic clusters (structure 3 in Figure 4.17).

The methoxyl groups undergo a shielding due to the formation of aromatic clusters in the char (Figure 4.15(a)) [BAR 07]. Hence, the δ^1H chemical shift from methoxyl groups (2.9–3ppm) can be distinguished from other δ^1H chemical shifts from carbohydrates (3.2–3.6 ppm) (Figure 4.16(c)). This finding is of tremendous importance to understand the chemistry of lignin and carbohydrates pyrolysis in the network and is enabled by our high magnetic field combined with the FSLG method. The formation of aromatic moieties from carbohydrates and lignin conversion is thus clearly distinguished.

The peaks at 146–147/6.0 ppm could correspond to furanic moieties cross-linked by the conversion of carbohydrates [BAC 09]. The furanic carbons correlate with aromatic and aliphatic protons (146–147/3.5 ppm) (Figure 4.17, structure 1, carbon 11). This 11 carbon signal is overlapped by the 3_4 ^{13}C signal, which correlates with aromatic ^1H from aromatic clusters and ^1H from methoxyl functions (147–146/6.0 ppm and 3.0 ppm).

The important peak at 132/6.0 ppm related to aromatic clusters [BAC 09] could be assigned to the aromatic carbons 3_5 formed from lignin [CHU 13] or from carbohydrates [ZAW 02]. Correlations at 132/2.9 ppm highlight the methoxyl groups remaining in the char from lignin.

Cross peaks at 127/6.0 ppm could be assigned to benzenic carbons from carbohydrates (structure 1 in Figure 4.17) because these carbons are correlated with aliphatic protons (3.6 ppm) from sugars.

The signal at 118/6.0 ppm (Figure 4.16) can be assigned to 2_3 carbons [BAC 09] (correlation with aromatic ^1H from furanic moieties). It can be observed that this signal is not correlated with any aliphatic or methoxyl proton, demonstrating the absence of aliphatic moieties in structure 2. The signal at 111/6.0 ppm could be assigned to 1_2 and 2_2 with a direct correlation with aromatic ^1H.

This 2D NMR method has been combined with a 1D ^{13}C NMR quantitative method. A specific cross polarization/magic angle spinning (CP/MAS) method has been developed and compared with direct polarization, which is quantitative but requires a very long accumulation time in the NMR [CAO 12].

The peaks for 1D NMR were assigned from literature [BAR 09, CAO 12] and deconvoluted by the DMFIT® software (from CNRS Orléans [MAS 02]). The 1D NMR method gives the evolution of the main chemical structure in *Miscanthus* and char as a function of the final temperature (Figure 4.18), and the 2D NMR method gives a qualitative outlook on the chemical moieties and connectivities between ^{13}C and ^{1}H in the network.

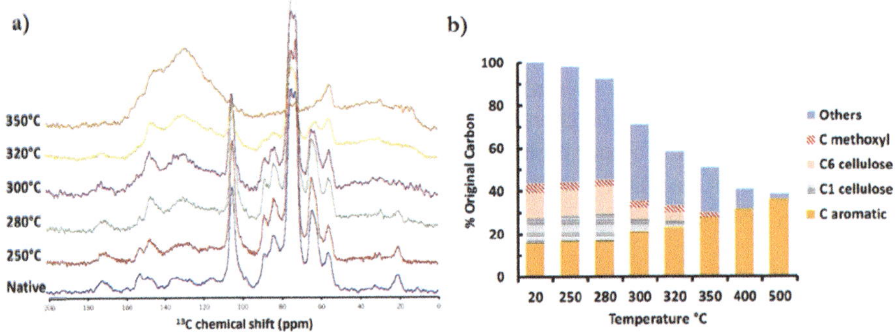

Figure 4.18. *(a) CP/MAS quantitative spectra obtained for different values of final pyrolysis temperature of Miscanthus (in fixed bed, 5 K/min) and (b) evolution of main carbonaceous moieties in char as a function of final temperature after deconvolution of the spectra. For a color version of the figure, see www.iste.co.uk/dufour/biomass.zip*

Figure 4.18 shows that the main chemical modifications in char occur between 280°C and 320°C at our heating rate (5 K/min) with devolatilization (carbons to gas phase) and cellulose conversion. New aromatic moieties are mainly formed by carbohydrates conversion at 300°C. Lignin depolymerization and xylan conversion are evidenced qualitatively at 250°C (and even lower temperatures) by 1D and 2D NMR. Only 40% carbon atoms originally present in *Miscanthus* remain at 500°C and mainly as aromatic carbons.

4.4. Molecular mechanisms of gas-phase reactions

4.4.1. *Introduction*

In section 4.3, we have seen how primary tar are devolatilized (as vapor) from biomass and char during primary pyrolysis. In many reactors, especially in high-temperature reactors such as gasification and combustion, primary tar then undergo secondary and tertiary reactions (Figure 4.19) depending on the thermal severity of gas-phase treatment (temperature and residence time of the gas/vapor phase). These reactions generate not only gas, which can be an important source of syngas production in gasification reactor, but also polycyclic aromatic hydrocarbons (PAH) and benzene (major tertiary tar). These tertiary products are pollutants in combustion smoke and in syngas and should be reduced. For this reason, it is important to better understand their pathways of formation and conversion.

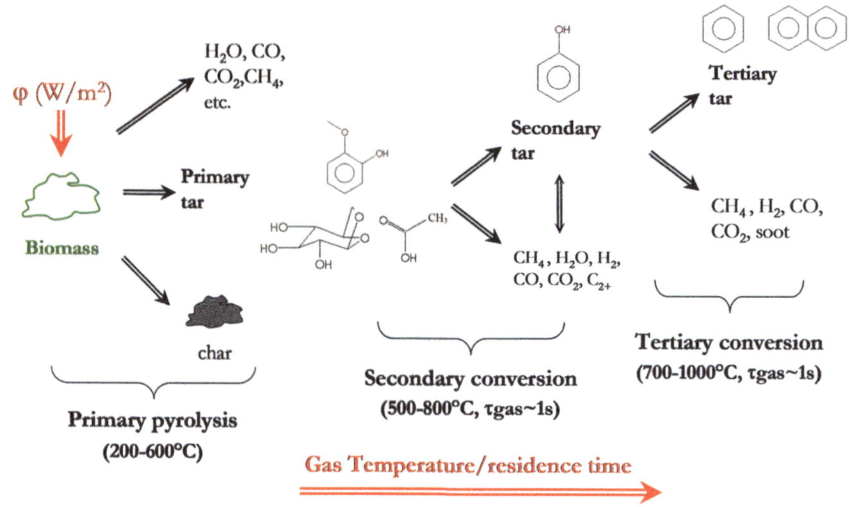

Figure 4.19. *Simplified scheme depicting gas-phase secondary and tertiary tar formation, adapted from [EVA 87]*

4.4.2. *Gas-phase conversion of model compounds*

The mechanisms of gas-phase conversion of primary tars have been little studied [FRI 01, SHI 01, SCH 11, LED 13]. Moreover, there are very few works that combine well-controlled experiments (controlled temperature and residence time of tar), tar quantitative analysis and elementary modeling. In our laboratory, we had the opportunity to combine skills from the combustion kinetics and biomass groups to study the gas-phase conversion of tar. Our laboratory develops detailed chemical mechanisms with thousands of elementary reactions [HAK 09]. These mechanisms are predictive on a broader range of conditions (temperature and residence time) than the lumped kinetic mechanisms.

The conversion of model primary tar compounds (anisole, guaiacol and 5-methylfurfural) was studied in a jet stirred reactor (JSR) as a function of temperature and O_2 concentration. These three species were chosen as a good surrogate of primary pyrolysis tar from lignin (anisole and guaiacol) and from cellulose (5-methylfurfural) [NOW 14].

A JSR (Figure 4.20) is the perfect reactor to determine the kinetics of gas-phase reactions because it can be considered as an ideal zero-dimensional perfectly stirred reactor with homogeneous temperature and concentration of species in its whole volume. This unique feature is obtained by the good mixing of the gas phase provided by the four turbulent jets in the injection cross.

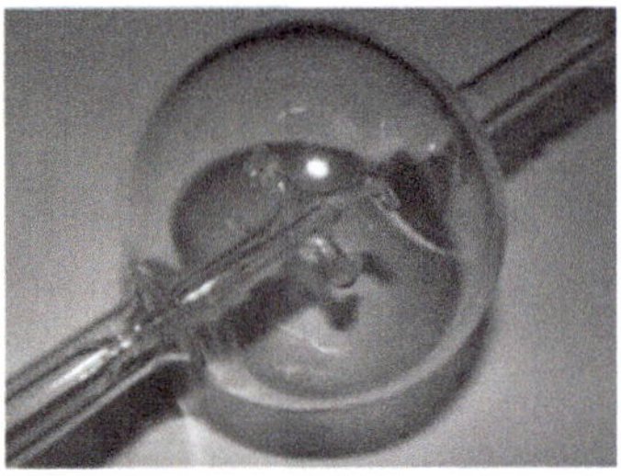

Figure 4.20. *Jet stirred reactor in silica used for the study of model tar gas-phase conversion at CNRS-Nancy [NOW 14]*

The outlet of the JSR is connected to various gas chromatography systems (including GC/MS-FID) for the quantification of the minor and major products (commonly more than 30 products).

Figure 4.21 depicts the conversion of the three compounds as a function of temperature under pyrolysis and oxidation conditions.

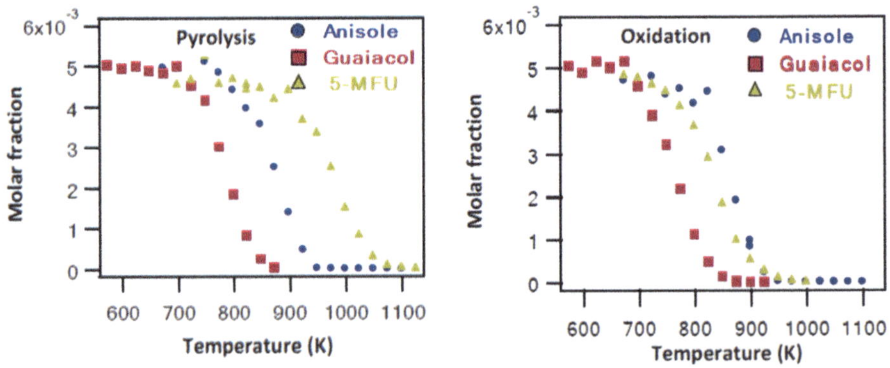

Figure 4.21. *Evolution of the mole fractions of anisole, guaiacol and 5-methylfurfural (5-MFU) under pyrolysis and oxidation (O_2) conditions as a function of temperature (residence time of 2 s)*

Detailed mechanisms were developed to predict the molar fractions of the major and minor products. These detailed mechanisms help explain (partially) the formation of PAH during tar gas-phase conversion. PAH are mainly formed from phenoxy radicals that generate cyclopentadienyl radicals (Figure 4.22). The hydrogen-abstraction-carbon-addition (HACA) mechanism (PAH formation based on C_2 or C_3 species) [WAN 97] could also occur but rather at higher temperatures [VAS 09].

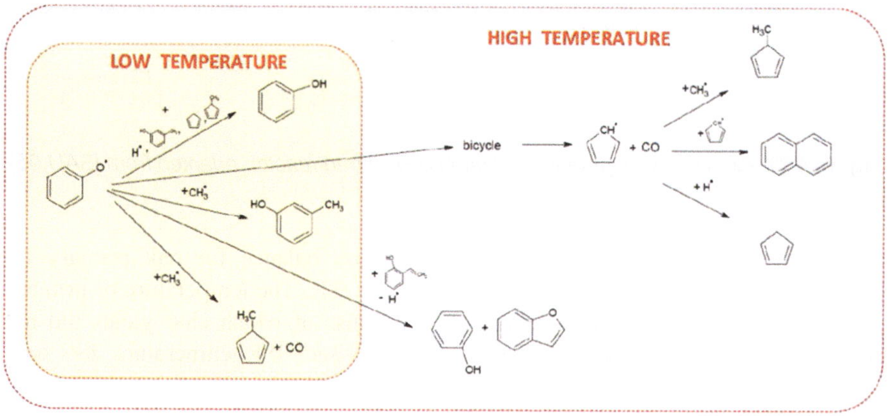

Figure 4.22. *Main reaction pathways for phenoxy radical conversion at about 1,000 K*

4.4.3. *Gas-phase conversion of real primary tar*

The gas-phase conversion of real primary tar has also been studied in our laboratory in a JSR (Figure 4.23) [BAU 05]. The primary pyrolysis is conducted in a tubular reactor (in a "spoon," first furnace) and the as-produced volatiles are directly injected in the JSR positioned in a second furnace. The temperature of the JSR can thus be modified without modifying the temperature of biomass pyrolysis. This setup illustrates well the decoupling methodology previously introduced in this book (section 4.1).

The mass balances (gas, tar, water, char) and the composition of gas and main tars were studied for cellulose, xylan, lignin and oak as a function of the temperature of the JSR.

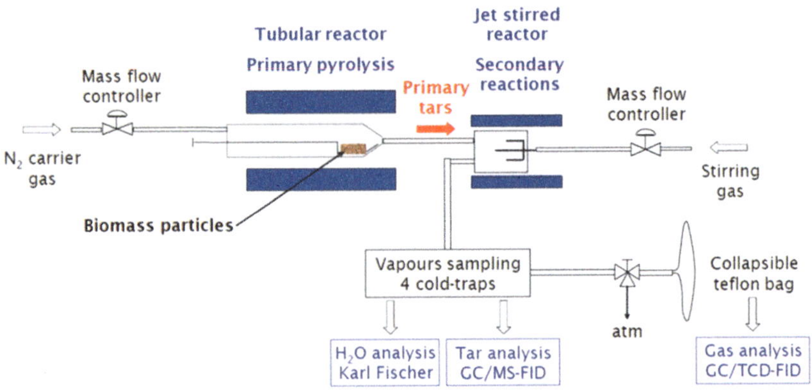

Figure 4.23. *Conversion of primary real tar in a jet stirred reactor, adapted from [BAU 05]*

Figure 4.24 shows the evolution of the mass balance for oak primary tar conversion as a function of the temperature of the JSR. The temperature of primary pyrolysis reactor was set constant at 540°C because of which char yields did not change at various JSR temperatures (char does not "see" JSR temperature, JSR only used for gas-phase conversion).

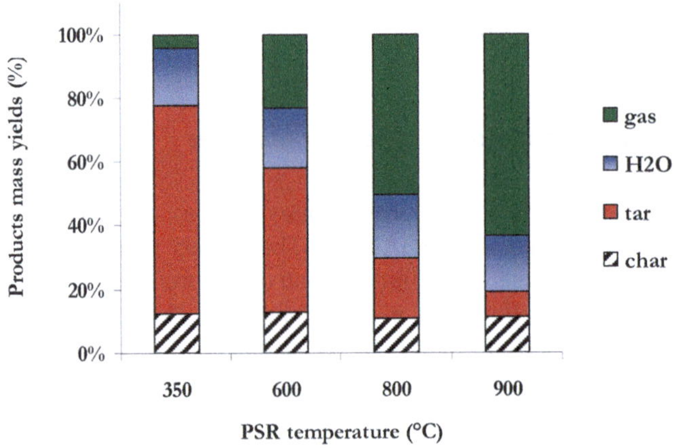

Figure 4.24. *Repartition between tar, gas and char as a function of the temperature of JSR: high temperature of gas-phase JSR results in tar conversion to gas. Gas residence time of 0.4 s in the JSR. For a color version of the figure, see www.iste.co.uk/dufour/biomass.zip*

Figure 4.25 highlights the huge dispersion of kinetic constants (k) determined in the literature for tar gas-phase conversion. This can be explained by different tar compositions (primary or secondary tar do not have the same reactivity) and different reactors used to derive these kinetic constants.

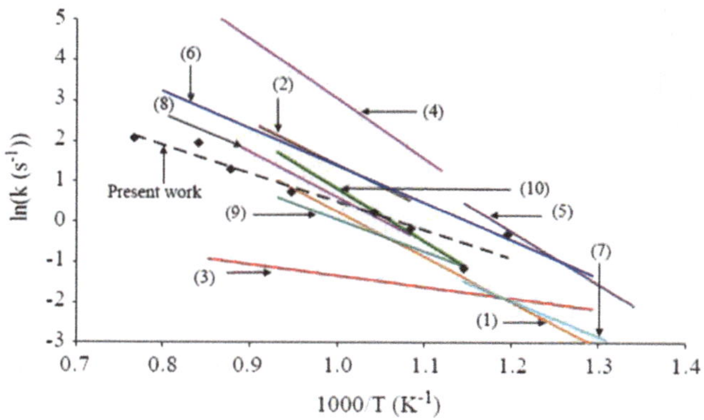

Figure 4.25. *Important dispersion in kinetic constant (k) determined in the literature for tar gas-phase conversion [BAU 05]. For a color version of the figure, see www.iste.co.uk/dufour/biomass.zip*

Primary tar (< 600°C, τ ~ 1s)

Compounds produced from cellulose and hemicellulose primary pyrolysis (levoglucosane, hydroxyacetaldehyde, furfural, furane,...) and from lignin (methoxyphenols).

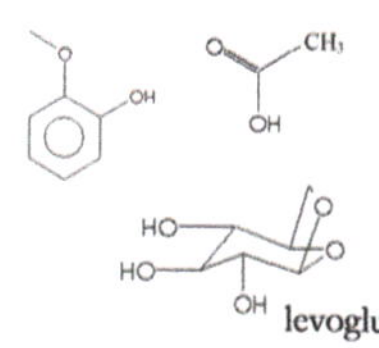

levoglucosan

Secondary (~ 500-800°C)

Olefins (=alcenes double bound C=C) and phenolics.

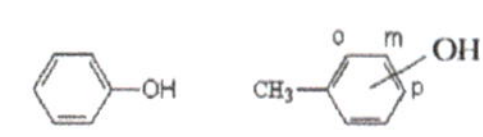

phenol cresols

Tertiary tar (> 700°C)

Mono and poly aromatics hydrocarbons (PAH) substituted with a methyl function (methyl naphthalene, toluene, etc.) or not (benzene, naphthalene, anthracene, pyrene, etc.).

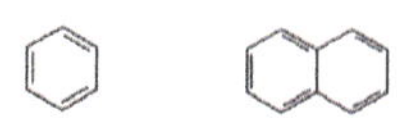

Benzene Naphthalene

Soot precursors.

Figure 4.26. *Composition of primary, secondary and tertiary tar, adapted from [EVA 87]*

Figure 4.26 sums up the composition of primary, secondary and tertiary tar (tar "maturation").

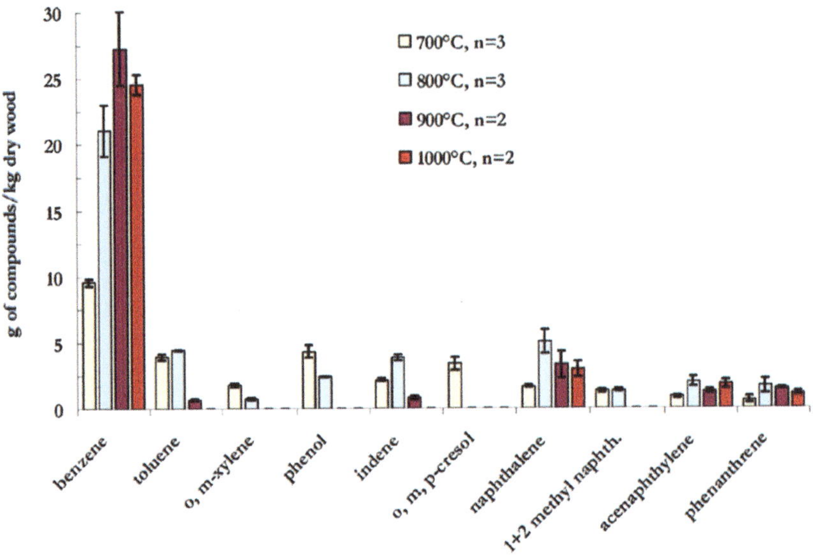

Figure 4.27. *Major aromatic tars produced during the pyrolysis of wood as a function of reactor temperature (n, number of experiments) [DUF 11]. For a color version of the figure, see www.iste.co.uk/dufour/biomass.zip*

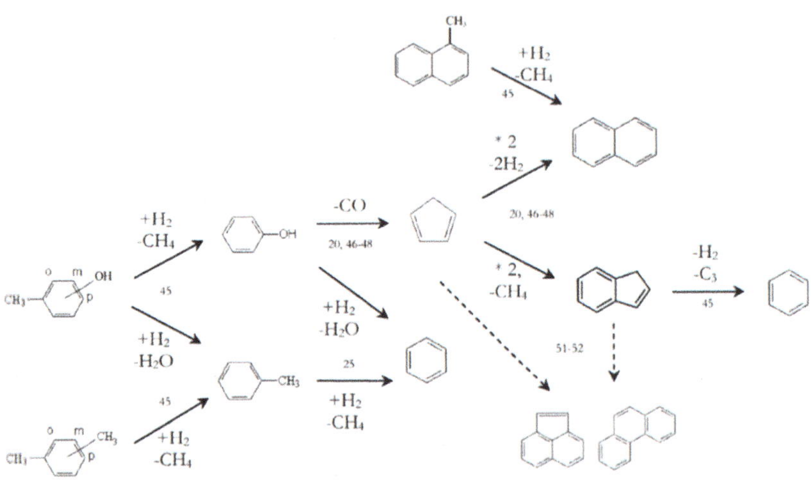

Figure 4.28. *Simplified mechanism of gas-phase conversion of secondary tars into tertiary tars*

The gas-phase conversion of secondary tar has been studied in a similar tubular setup (with a "spoon" for the injection of biomass) by our group [DUF 11a]. The major aromatic compounds produced by the high-temperature gas-phase reaction (700–1000°C) were carefully quantified by GC/MS and deuterated internal standards (Figure 4.27). It can be seen that naphthalene and other PAH are formed at 700°C (for a gas residence time of 2.7 s) confirming the finding from model tar compounds that the HACA mechanism (based on C_2 and C_3) is a minor pathway for PAH formation under these conditions. PAH are most probably formed from the conversion of phenoxy radicals. A simplified mechanism of PAH formation in a wood pyrolysis gas is proposed in Figure 4.28.

We have also shown that methane and ethylene could be used as indicators of the gas-phase "maturation" of tars (Figure 4.29). These relations should be checked for other conditions. They would not be suitable for reactors with important catalytic conversion of tars but they could be valid for dual fluidized-bed reactors.

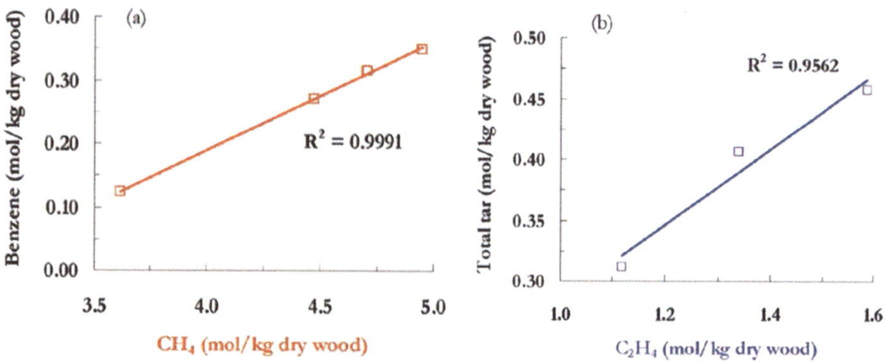

Figure 4.29. *Relations between (a) benzene and methane and (b) total tar and ethylene [DUF 11]*

4.4.4. *Gas-phase conversion of methane in a syngas*

The gas-phase conversion of methane in a reconstituted syngas was investigated by our group in a tubular reactor (in collaboration with CEA, Grenoble, France) at 130 kPa, for a gas residence time of 2 s and as a function of temperature (1000–1400°C), CH_4 (7, 14%), H_2 (16, 32%) and H_2O (15, 25, 30%) initial mole fractions [DUF 09b]. H_2 showed an inhibiting effect on CH_4 conversion whereas H_2O had few effects. Three detailed elementary mechanisms were used to predict methane conversion rate and to identify the key reaction pathways. The mechanism developed by our laboratory was shown to predict all the experimental results very well. Flow rate analyses showed that carbon oxidation occurs mainly by addition of OH radicals on C_2 compounds. A new mechanism for methane conversion in a syngas has been derived (Figure 4.30). OH radicals are mainly produced by CO_2 ($CO_2 + H = CO + OH$). The inhibiting role of H_2 on CH_4 conversion is explained by a competition between the OH radicals consumption channels ($H_2 + OH = H_2O + H$). The formation of soot has been explained by the competition between thermal conversion of methane and reformation of unsaturated C_2.

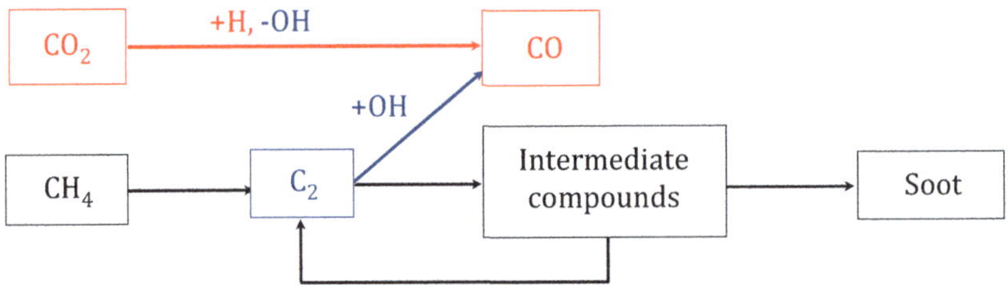

Figure 4.30. *Mechanism of methane conversion in a syngas [DUF 09]. The oxidation of carbons occurs by the addition of OH radicals, produced by CO_2 (and not by H_2O), on C_2 compounds*

4.5. Molecular mechanisms of catalytic reactions

4.5.1. *Catalytic fast pyrolysis of biomass*

The catalytic fast pyrolysis of biomass has been paid recently a tremendous interest to produce green aromatics and olefins from biomass. In this process, biomass is pyrolyzed in a fluidized bed in the presence of an acid catalyst such as zeolite. The overall chemistry is presented in Figure 4.31. Primary tar (at the vapor phase) interacts with the surface of zeolite to form a hydrocarbon pool of small molecules, which further produced aromatics or olefins in similar mechanism than the methanol to olefins process. The aromatic yield has been shown to mainly depend on the pore size of zeolite (Figure 4.32) [JAE 11]. Medium-pore zeolites with moderate internal pore space and steric hindrance (ZSM-5 and ZSM-11) gave the highest aromatic yield and the least amount of coke. The main drawback of this technology is the important production of coke, which results in a loss of carbon (Figure 4.33). Zeolite has to be regenerated very fast by oxidation in dual fluidized-bed reactors such as the fluid catalytic cracking process in petroleum refinery.

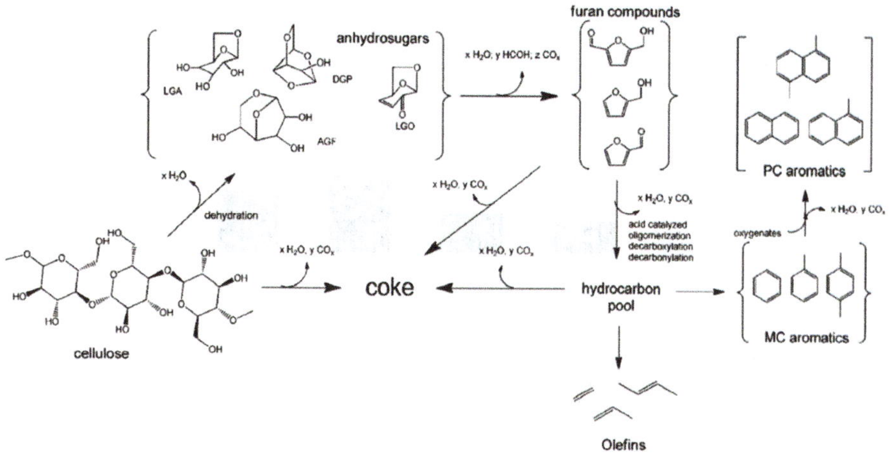

Figure 4.31. *Overall reaction chemistry for the catalytic fast pyrolysis of cellulose [CAR 10]*

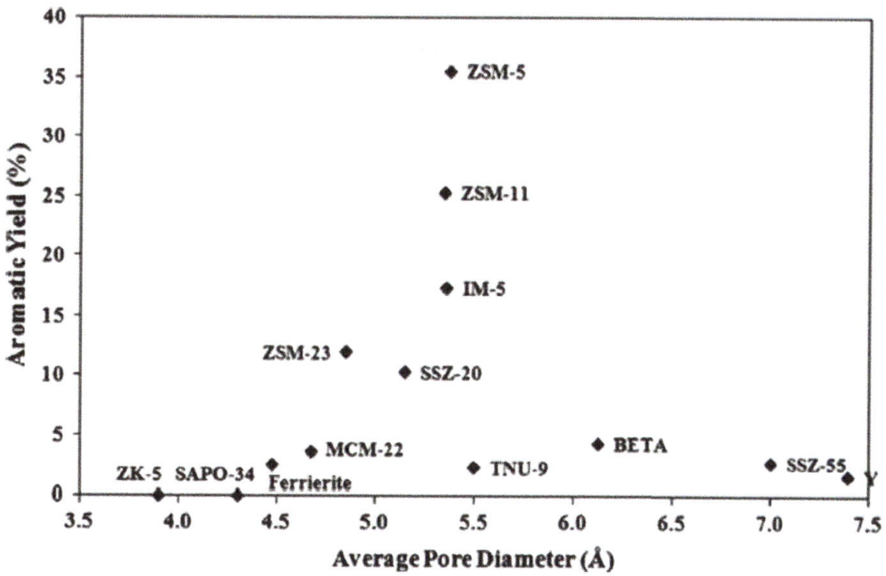

Figure 4.32. *Aromatic yields as a function of average pore diameter for different zeolites for the catalytic fast pyrolysis of glucose [JAE 11]*

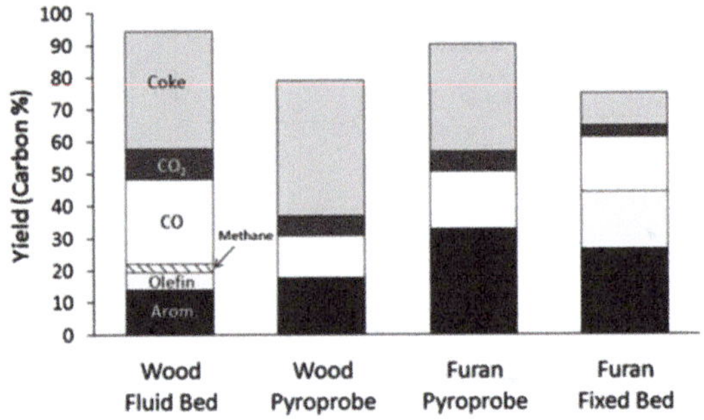

Figure 4.33. *Comparison in carbon yields in main products for different reactors (fluidized bed or milligram-scale pyrolysis in a pyroprobe) [CAR 10]*

4.5.2. *Catalytic reforming of tar to produce a syngas*

A large variety of catalysts has been tested for more than 50 years to convert tar in gasification systems. Catalytic conversion of tar has been reviewed several times [SUT 01, ABU 04, YUN 09, BUL 11].

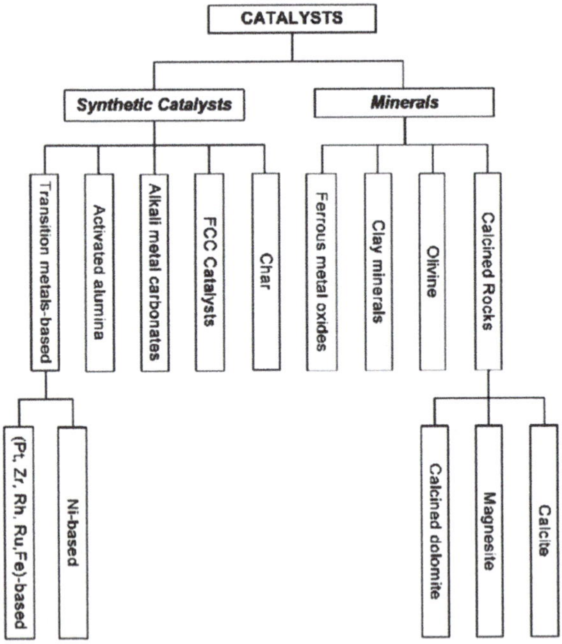

Figure 4.34. *Classification of catalysts tested for tar conversion [ABU 04]*

catalyst	advantages	disadvantages
calcined rocks	inexpensive and abundant attain high tar conversion ~95% conversion with dolomite often used as guard beds for expensive catalysts most popular for tar elimination	fragile materials and quickly eroded from fluidized beds
olivine	inexpensive high attrition resistance	lower catalytic activity than dolomite
clay minerals	inexpensive and abundant fewer disposal problems	lower catalytic activity than dolomite most natural clays do not support the high temperatures (800–850°C) needed for tar elimination (lose pore structure)
iron ores	inexpensive abundant	rapidly deactivated in the absence of hydrogen lower catalytic activity than dolomite
char	inexpensive natural production inside the gasifier high tar conversion comparable to dolomite	consumption because of gasification reactions
FCC	relatively inexpensive but not cheaper than the above more known about it from experience with FCC units	rapid deactivation by coke lower catalytic activity than dolomite
alkali-metal-based	natural production in the gasifier reduced ash-handling problems	particle agglomeration at high temperatures lower catalytic activity than dolomite
activated alumina	high tar conversion comparable to that of dolomite	rapid deactivation by coke
transition-metal-based	able to attain complete tar elimination at ~900°C increases the yield of CO_2 and H_2 Ni-based catalysts are 8–10 times more active than dolomite	rapid deactivation because of sulfur and high tar content in the feed relatively expensive

Table 4.2. *Advantages and disadvantages of different catalysts for tar conversion [ABU 04]*

The global targeted reactions are steam or CO_2 (dry) reforming of tar to produce CO and H_2:

steam reforming	$C_mH_n + nH_2O \leftrightarrow nCO + (n + \frac{m}{2})H_2$
dry reforming	$C_mH_n + nCO_2 \leftrightarrow 2nCO + (\frac{m}{2})H_2$
thermal cracking	$C_mH_n \rightarrow C^* + C_xH_y + gas$
hydrocracking or hydroreforming of tars	$C_mH_n + H_2 \leftrightarrow CO + H_2 + CH_4 + \cdots + coke$
water – gas shift reaction	$CO_{(g)} + H_2O_{(g)} \leftrightarrow CO_{2(g)} + H_{2(g)}$

The water–gas shift reaction is also an important reaction enhanced by the catalysts.

Catalysts can be classified as shown in Figure 4.34 [ABU 04].

Table 4.2 summarizes the main disadvantages and advantages of these different catalysts. Importantly, these catalysts should be cheap because they are always prone to deactivation by coke deposit or by ash coating [KIM 11].

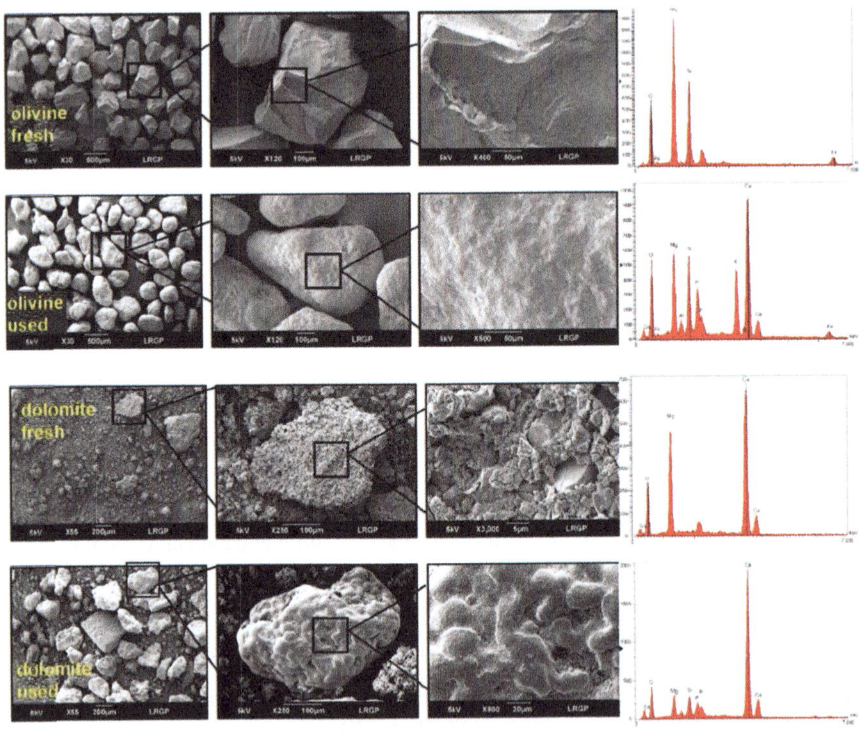

Figure 4.35. *SEM-EDX analysis of fresh and deactivated olivine and dolomite, which reveals ash coatings after using the catalysts in industrial air-dense fluidized-bed gasifier. For a color version of the figure, see www.iste.co.uk/dufour/biomass.zip*

Our group has conducted advanced analysis of dolomite and olivine deactivated in an industrial air-dense fluidized-bed gasifier. Figure 4.35 shows scanning electron microscopy (SEM) of fresh and deactivated catalysts analyzed in CNRS-Nancy. Ash coating is clearly evidenced by energy-dispersive X-ray (EDX) analysis. Ash coating modifies the surface chemistry of the catalysts.

Important developments of new catalysts have been undertaken for this specific application (tar conversion in fluidized bed) notably in the University of Strasbourg (France) [VIR 12].

Special materials have been also designed for combined biomass gasification and CO_2 chemical looping [DIF 11].

4.5.3. *Catalytic hydrodeoxygenation of bio-oils*

4.5.3.1. *Basics of bio-oils deoxygenation*

In the last decade, the deoxygenation of bio-oils has also been paid an increasing interest in order to produce biofuels and green chemicals from biomass [ELL 15a]. The concept is presented in Figure 4.36.

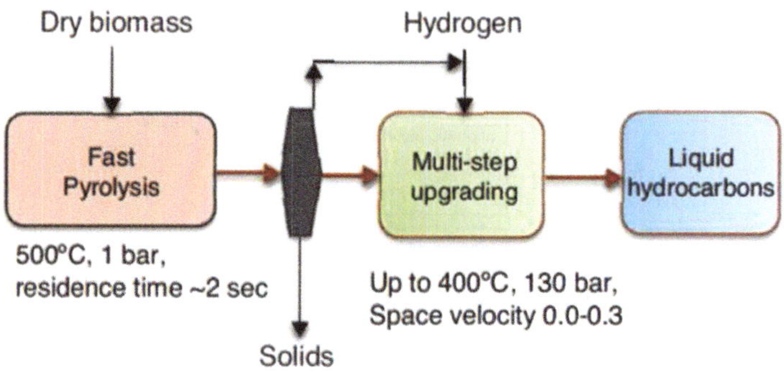

Figure 4.36. *Principle of hydrodeoxygenation of bio-oils for the production of hydrocarbons [ELL 15a]*

The catalyst should be selective to remove selectively the oxygen atoms from the various chemical moieties present in bio-oils to produce deoxygenated hydrocarbons. The main chemical pathways are presented in Figure 4.37.

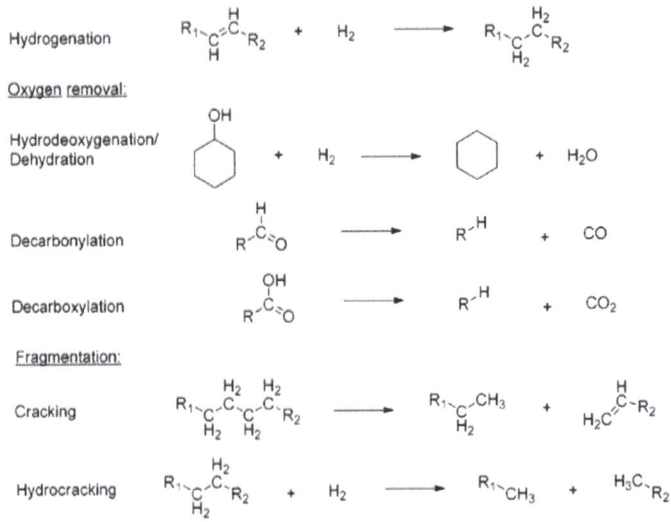

Figure 4.37. *Main reactions occurring during hydrotreatment of bio-oils [ZAC 14]*

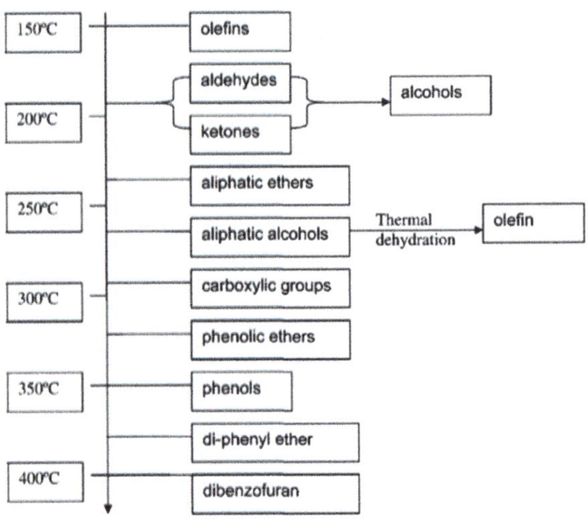

Figure 4.38. *Stability of different functional groups during catalytic hydrotreatment [CEN 95]*

To be more selective for the various compounds present in bio-oils that have different reactivities (Figures 4.38 and 4.39), the hydrotreatment should be staged (two or even three stages) with a first mild hydrotreatment (at a low temperature of about 150°C) to deoxygenate very reactive compounds (such as aldehydes) and to reduce coke formation and plugging of the catalytic reactor. A two-stage non-isothermal reactor configuration is currently used, which consists of a low-temperature stage (150–200°C) with RuS/C catalysts and a high-temperature stage (350–405°C) under 140 bar of H_2 [ZAC 14].

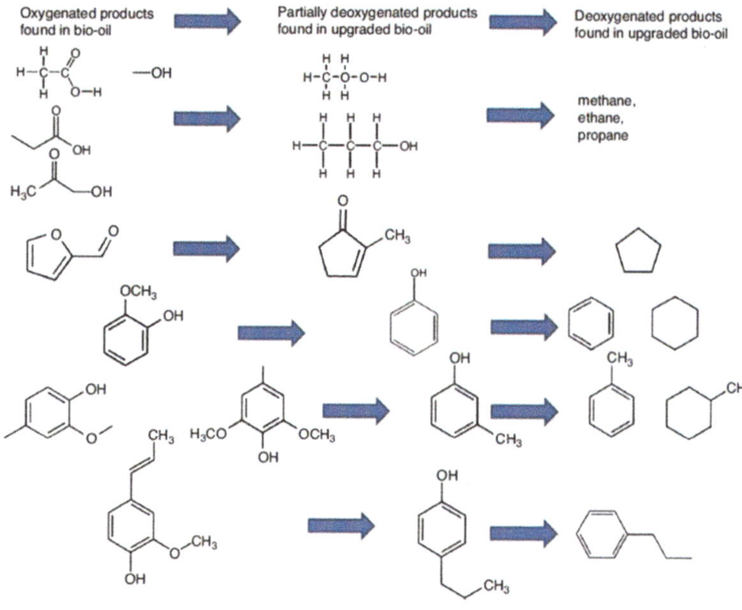

Figure 4.39. *Compounds found in bio-oil and the resulting hydrotreatment products*

4.5.3.2. *Catalytic hydrodeoxygenation of lignin primary tar to produce aromatic chemicals*

4.5.3.2.1. Catalytic hydrodeoxygenation of guaiacol as a surrogate of lignin primary tar

In the frame of lignin conversion into aromatic chemicals, our group has worked on the hydrodeoxygenation (HDO) of guaiacol. Guaiacol was chosen as a surrogate compound of lignin primary tar. Guaiacol is a minor but significant component in the very complex mixture of lignin bio-oils [NOW 10]. It has been chosen as a model compound because its elemental composition is close to the overall elemental composition of lignin bio-oils (see H/O/C ratios in Figure 4.40) and its hydroxyl and methoxyl functions on the aromatic ring are significant and key functions for aromatics species in lignin bio-oil.

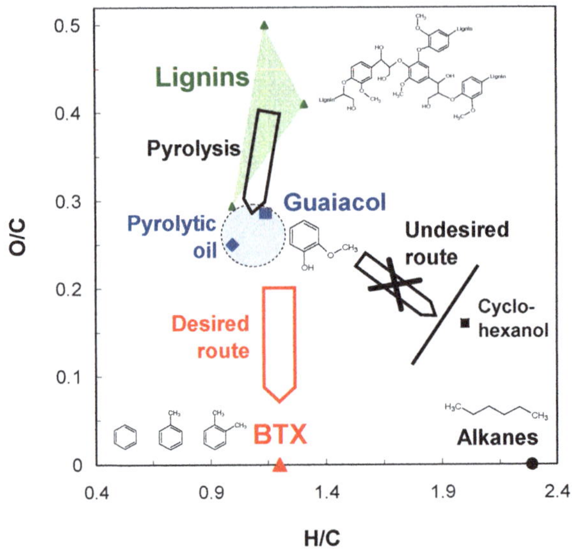

Figure 4.40. *Van Krevelen diagram depicting the desired selectivity route for the catalytic hydrodeoxygenation of lignin bio-oil into aromatics [OLC 12]. BTX, benzene, toluene and xylenes*

The catalyst of HDO should promote a selective hydrogenolysis of the $C_{aromatic}$–O bond to convert guaiacol into benzene, toluene and xylenes (BTX) but should not activate the hydrogenation of the aromatic ring.

Extensive work has been conducted on the HDO of surrogate compounds from lignin bio-oil using Ni, Mo, Co, Pt (etc.)-based catalysts [SHI 00, ROM 10, BAD 11, BUI 11, ZHA 11, ZHU 11]. Transition (Fe, Co, Ni) and precious metals (Pt, Pd, Ru, Rh, Ir, etc.) are known to catalyze hydrogenation [EMM 43], but unfortunately most of them also catalyze benzene hydrogenation in the temperature range of 200–400°C [YOO 83]. Emmett and Skau [EMM 43] detected no activity for benzene hydrogenation at 400°C working with Fe. Lately, Yoon and Vannice [YOO 83] measured a relatively low activity of Fe for benzene hydrogenation compared to other metals (Ni, Co). Fe is a trade-off between activity and selectivity. For this is reason, we have chosen iron as the active phase in the guaiacol HDO, expecting minimum aromatic ring hydrogenation.

Metal–support interactions and support acidity play a crucial role in complex chemistry of transition metal–supported catalysts. For HDO reaction on supported Mo catalysts, alumina gave higher activity but rapid deactivation due to coke deposit [CEN 95]. In contrast, the use of the less acidic support as active carbon or silica led to lower activity but a better selectivity in HDO products [CEN 95, POP 10]. We concluded that silica or activated carbon (AC) would be good iron supports due to their low acidity [OLC 12].

Our work has focused on the gas-phase HDO of guaiacol to BTX using Fe/SiO_2 and Fe/AC, at atmospheric pressure and at the temperature of interest (400–500°C) for coupling the HDO catalytic reactor to the pyrolysis reactor (450–550°C).

The catalysts before and after reaction were characterized by N_2 sorption, X-ray powder diffraction (XRD), Mössbauer spectroscopy, transmission electron microscopy (TEM) and temperature-programmed oxidation (TPO) (for quantification of coke deposit) [OLC 13].

The effects of space time, H_2 partial pressure, temperature, gas composition (systematic effect of CO, CO_2, H_2 and H_2O), iron loads and support (silica and AC) on product selectivity and catalyst deactivation have been studied. The speciation of the iron phase (metallic (α-Fe), carbide (Fe_5C_2), oxide (Fe_3O_4), and super-paramagnetic) in spent catalysts was revealed by XRD and Mössbauer spectroscopy as a function of gas composition. When all the gases except methane (guaiacol + H_2 + CO + CO_2 + H_2O) were simultaneously in the feed stream, the conditions were still sufficiently reducing to maintain the activity of the catalyst (66% benzene and toluene carbon yield, 7.5 g_{cat} h/g_{gua}) [OLC 13].

A mechanism of guaiacol HDO over Fe/SiO_2 has been proposed (Figure 4.41) and the kinetic rates of the main reactions were derived (Figure 4.42).

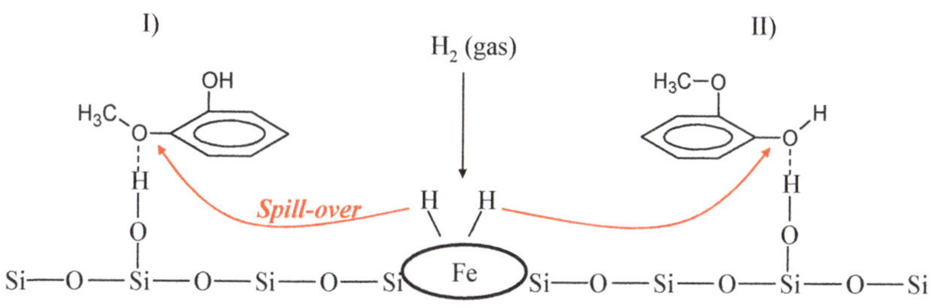

Figure 4.41. *Possible reaction mechanism of guaiacol conversion into aromatic hydrocarbons by HDO over Fe/SiO_2. The adsorption of guaiacol was proposed by Popov et al. [POP 10, OLC 12]*

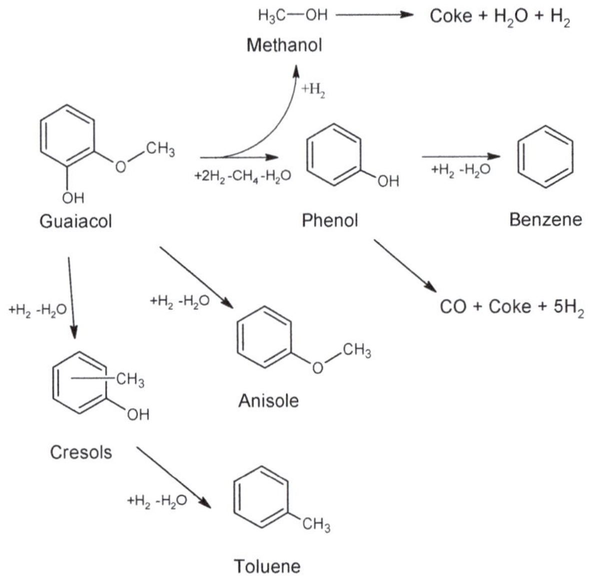

Figure 4.42. *Main pathways for the HDO of guaiacol over Fe/SiO2 [OLC 13]*

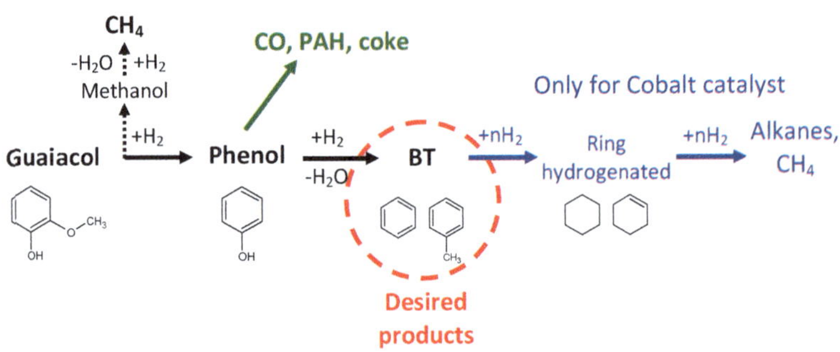

Figure 4.43. *Simplified chemical pathways for HDO of guaiacol over Fe/SiO2 and Co/SiO2. PAH, polycyclic aromatic hydrocarbons*

We have conducted kinetic studies including the determination of deactivation laws at low and high conversion of guaiacol. The kinetic studies at initial and low conversions are suitable to analyze the fundamental kinetics on fresh catalysts, but they are not of practical interest to predict the rates of desired product formation at a significant yield and on a partially deactivated catalyst. For this reason, the experiments were conducted with a broad range of catalyst mass, including high guaiacol and phenol conversions. Such experiments are needed to demonstrate that the catalysts are really selective for benzene and toluene productions in a significant yield. Indeed, at high conversions, some catalysts could promote the hydrogenation or cracking of aromatic rings, becoming not selective for benzene production (such as cobalt-based catalyst, see Figure 4.43).

This does not happen with iron because it preserves the aromatic ring. We have proposed to plot the selectivity in benzene and toluene yield as a function of the $C_{aromatic}$–O bond cleavage (Figure 4.44). Such a diagram has been used (and should be used) to compare catalysts activity and selectivity for similar operating conditions. Iron was shown to be very selective for BTX production. It is less active than other catalysts from literature but much cheaper and more selective, so a bigger mass of catalyst can be used.

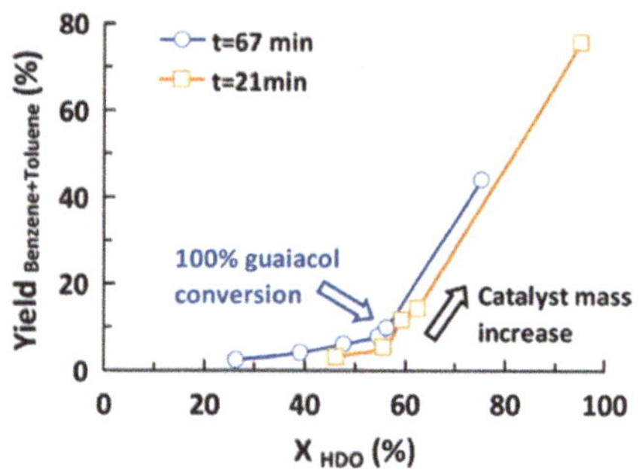

Figure 4.44. *Yield in benzene and toluene as a function of $C_{aromatic}$–O bond cleavage (X_{HDO}) and of the mass of catalyst used and for two times on stream (21 and 67 min) [OLC 13]. For one time on stream, each point corresponds to different catalyst masses. The selectivity in benzene and toluene should be proven at high HDO conversion, in such a diagram, to demonstrate that the catalysts are really selective for benzene and toluene productions in a significant yield. A very good selectivity is obtained with iron supported over silica*

Finally, an important feature of this catalyst is its capability to be used with only 0.2 bar of H_2 partial pressure. We have shown that its selectivity for BTX production between 0.2 and 1 bar of H_2 (20–100 %mol. H_2 at atm. pressure) is similar.

4.5.3.2.2. Catalytic hydrodeoxygenation of lignin real vapors

The HDO on real lignin pyrolysis vapors was also studied by our group [OLC 13]. The vapors were produced by the injection of lignin in a preheated tubular reactor (as in Figure 4.23). The volatiles were directly injected in the catalytic fixed bed. The fixed bed was composed of four stacks (separated by quartz wool plug) to investigate the profile of coke deposit along with the length of the fixed bed. SiO_2

and AC were compared as support for iron. Successive loads of lignin were injected on the same catalytic bed to assess the deactivation of the catalyst. The vapours were collected in condensers and diluted in methanol (see figure 4.45).

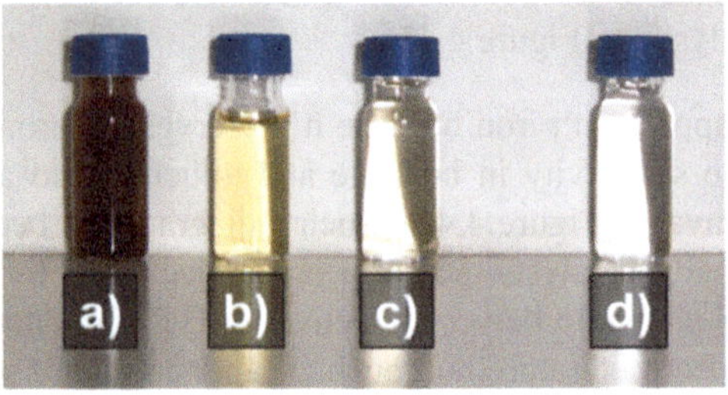

Figure 4.45. *Picture of lignin bio-oils dissolved in 15 ml methanol, after one load of lignin: (a) without catalytic treatment (empty fixed-bed reactor at 673 K), (b) using 2 g of 15% Fe/SiO₂, (c) using 4 g of 15% Fe/SiO₂ and (d) using 2 g of 10% Fe/AC*

The selectivity of the catalyst on GC-analyzed compounds is plotted in a Van Krevelen diagram. The deoxygenation is very selective (Figure 4.46).

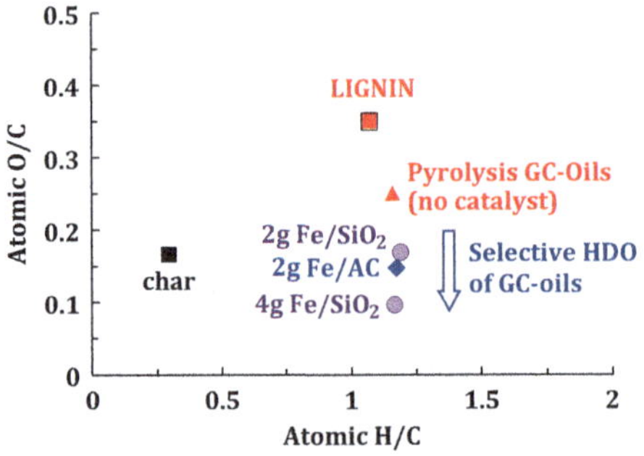

Figure 4.46. *Van Krevelen diagram of lignin, char (based on elementary analysis) and GC-analyzed oils (based on GC*GC advanced quantification) obtained from HDO of lignin pyrolysis vapors on iron-based catalysts. A very selective HDO is revealed: O/C atomic ratio is reduced without affecting H/C ratio*

The overall atomic ratio does depict the global selectivity of the hydrotreatment but not the high diversity of the compounds. Indeed, each point depicted in Figure 4.46 accounts for the contribution of tens of GC-analyzed species. The hundreds of compounds analyzed by GC*GC (heart-cutting)/MS-FID and petroleomic are presented in Figure 4.47.

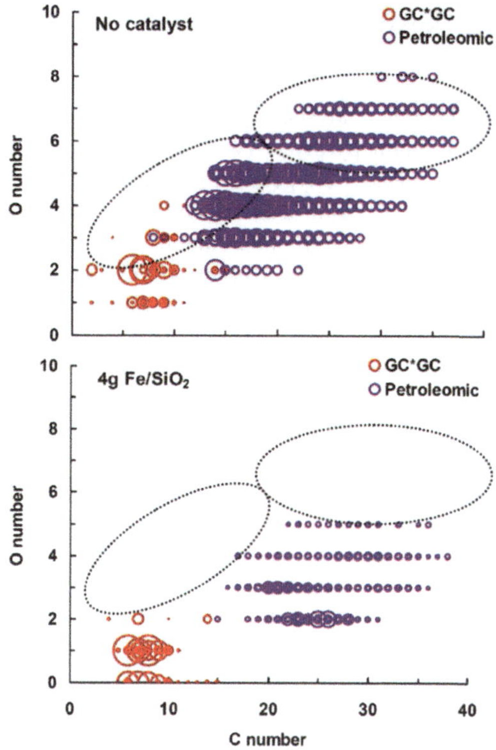

Figure 4.47. *Compound class (C and O numbers) determined by GC*GC and petroleomic analysis (LDI) as a function of weight yield (for GC) and relative abundance for petroleomic, for lignin pyrolysis without catalyst and 4 g/FeSiO$_2$ treatment [OLC 13]. The area of circles is function of weight yield for species analyzed by GC and relative abundance for petroleomic and cannot be compared between the two analytical methods. It can be compared only between the two experimental conditions and for the same analytical method. For a color version of the figure, see www.iste.co.uk/dufour/biomass.zip*

Some zones of interest are depicted in Figure 4.47. During the catalytic treatment, GC-oils (in red in Figure 4.47) are deoxygenated without an increase in the carbon number (little growth). Light species, such as acetic acid at C_2 and O_2, are converted by the catalytic treatment. O_2 class molecules are almost completely converted into O_1 and O_0 species. An interesting feature of the iron-based catalysts is their ability to preserve the aromatic rings and also to promote alkylation. The alkylation is well depicted by this chart (increase in C_8 species). Alkylation could result from the pool of methyl species at the surface of the catalysts, such as methanol and radicals. Deoxygenation and alkylation reactions are achieved without condensation of the C_6–C_8 species.

Regarding the petroleomic analysis, Figure 4.47 shows that the main species are O_4–O_5 and C_{13}–C_{20} for lignin pyrolysis (without catalyst), but after the catalytic treatment, heavy species are less oxygenated (mainly O_2 and O_3 classes) but with higher-carbon-number classes (C_{20} to C_{30}). Heavier lignin oligomers are more prone to growth and condensation than monoaromatic species analyzed by GC. The heaviest species (with more than five oxygen atoms) are completely trapped or converted by the catalytic bed (zones highlighted in Figure 4.47) [OLC 13].

The heavy species are most probably poorly converted into GC-oils otherwise the weight yields in GC-oils would have increased instead of decreasing after the catalytic treatment. They are mainly converted into coke deposit.

Coke deposit (Figure 4.48) was analyzed for the four stacks by TPO [OLC 13].

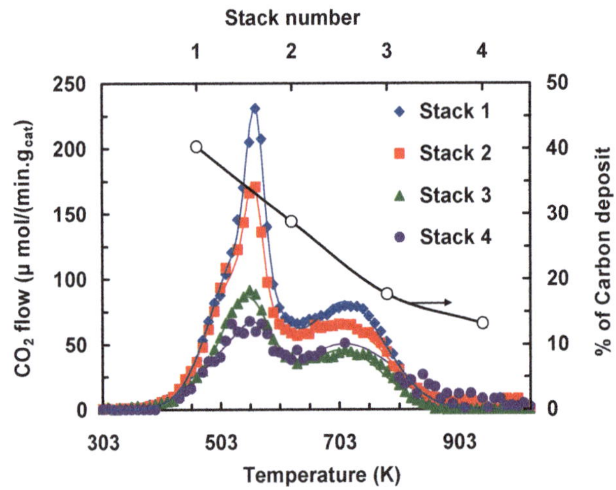

Figure 4.48. *TPO profiles of catalyst stacks after four loads of lignin. For a color version of the figure, see www.iste.co.uk/dufour/biomass.zip*

All TPO profiles show a shoulder at 500–540 K and two important peaks at 550–560 K and 706–714 K (Figure 4.48). They are completely different than those from guaiacol HDO used as a model compound. The shoulder (500–540 K) and low-temperature peak (550–560 K) could be produced by the gas-phase conversion of heavy oligomers from lignin pyrolysis [DAV 07, VAL 10, VAL 12]. This carbonaceous deposit is called "thermal coke." Valle *et al.* [VAL 12] studied the nature of carbon deposit from wood bio-oil/methanol cracking on Ni-doped H-ZSM5. They noticed that "thermal coke" (low-temperature TPO peak) may be originated by acid-catalyzed polymerization of lignin-derived tar. Deactivation by lignin-derived "thermal coke" was also shown in similar studies [AHO 10, VAL 12].

About 50%wt. of total carbon deposit accounts for this low-temperature peak for all stacks. The important decrease in this first peak along the fixed-bed length (Figure 4.48) shows that the catalytic fixed bed acts as "a filter" for heavy lignin oligomers. For this reason, low-cost catalysts such as Fe/SiO_2 should be looked for. Furthermore, process integration by coupling filters, staged condensation and catalytic conversion could help to increase lifetime and selectivity of catalysts [MIH 11].

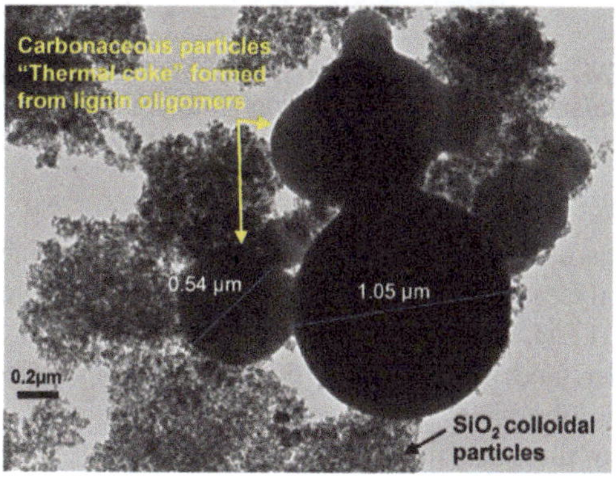

Figure 4.49. *TEM micrograph of Fe/SiO_2 catalyst used for the HDO of real lignin pyrolysis vapors (first stack) highlighting "thermal coke" formed from the gas-phase growth of lignin oligomers*

The first stack of Fe/SiO_2 catalyst was analyzed using TEM (Figure 4.49). For sake of comparison, a high-resolution TEM (HRTEM) image of the same catalyst exposed to H_2-guaiacol is shown in Figure 4.50 depicting the "catalytic coke" formed in the vicinity of iron particles. Carbonaceous particles, rich in carbon and oxygen (analyzed by EDX), with round well-defined forms were observed (Figure 4.49). These particles are probably formed by lignin oligomers (at the vapor or aerosols form) that grow into solid particles at reaction temperature (673 K) similar to the formation of "soot" from PAH in combustion processes. Considering the size of these carbonaceous particles (300–1100 nm diameters), they are probably formed from the gas-phase conversion of lignin oligomers with very minor interactions over the catalyst surface.

N_2 sorption analysis showed that coke deposit completely plugs the micropores in Fe/AC but not the mesopores (10–20 nm) of Fe/SiO_2.

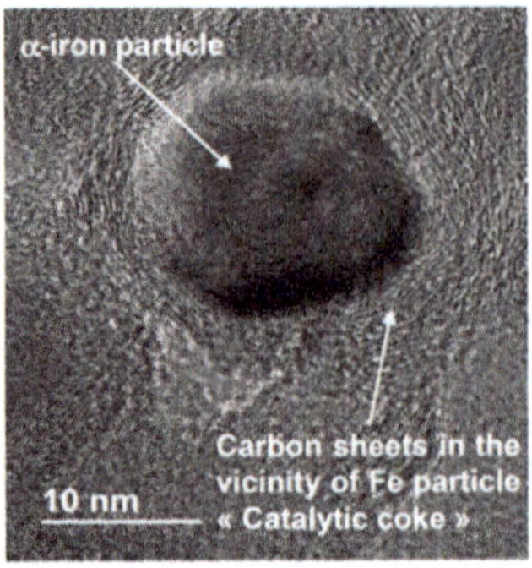

Figure 4.50. *HRTEM micrograph of an iron particle in Fe/SiO2 catalyst used for HDO of guaiacol at 400°C highlighting "catalytic coke" formed from phenols interactions with the catalyst surface*

Figure 4.51 sums up the main findings of this work, obtained with the various complementary analyses of bio-oils (by GC*GC and petroleomic) and of the catalysts (by TPO, TEM, N_2 sorption, XRD Mössbauer) [OLC 13]. More work is still needed to understand the formation of the two types of coke, especially the mechanism of lignin oligomers growth and the thermal annealing of coke deposits (at the temperature of the catalyst).

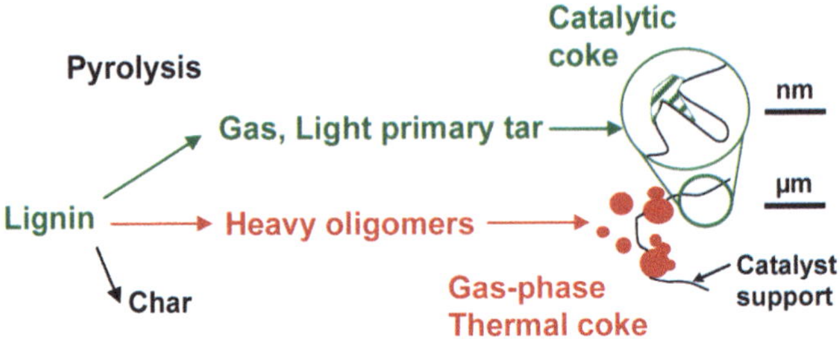

Figure 4.51. *Simplified mechanism of coke formation during the lignin pyrolysis vapors HDO over iron-based catalysts. The conditions of catalytic coke deposit, which plugs the pores, depend on the competition between reaction and diffusion rates in the pores [OLC 13]*

The main challenge for the HDO of lignin pyrolysis products, and also for biomass pyrolysis, is to convert the heavy lignin oligomers into fuels or valuable aromatics such as BTX or phenol. New catalytic systems have still to be looked for.

Some regeneration tests (by air oxidation of coke followed by H_2 reduction of iron) were also conducted. Iron-based catalysts have shown promising results but more experiments are still needed with a constant source of tar production (as in fluidized beds or screw reactor), with longer time on stream and many deactivation/regeneration cycles.

Conclusion

Example of an integrated research combining different scales of investigation

Figure 1 presents a simplified diagram describing a multiscale approach to improve the biomass conversion processes. This methodology is currently developed in our laboratory to improve the selectivity and to reduce the environmental footprint of thermochemical processes.

It is important to keep in mind the applied (industrial or social) aspects of a research. The studies of chains and processes point out process units with low energy efficiency and/or high environmental impact, and stress the improvement of reactors and the need of more fundamental knowledge (Figure 4.1). Conversely, the work at molecular and mesoscopic levels gives insight into the mechanisms of biomass conversion and is used to design reactors and to further improve processes.

The main findings from our group on conversion of lignin into aromatics and previously presented in the chapters on process and meso- scales (Chapters 1 and 3) are recalled in Figure 1.

At the process scale, the Aspen Plus® model of the lignin to benzene, toluene and xylenes (BTX) process has shown the detrimental effect of the carrier gas for process intensification. An important carrier gas flow rate must be used for pyrolysis in fluidized bed so the pyrolysis vapors will be diluted, and it will lead to higher investment costs for heat exchangers (heat recovery on carrier gas), higher electricity consumption of blowers and a bigger absorption column (higher "wash oil flow") for recovery of BTX vapors. The carrier gas could be recycled [DEW 12]. However, if the clean gas is recycled without CO_2 separation, the CO_2 partial pressure would increase in the pyrolysis and catalytic reactors, and the gas mixture could become an oxidizing mixture that will not reduce the catalyst. Fe/SiO_2

(or other reduced metals) would be transformed into Fe_3O_4/SiO_2 and its activity would be decreased. It would be also advantageous to decrease the fluidization gas flow rate [GEL 73] as much as possible or other technologies of pyrolysis reactors with a lower need in carrier gas flow rate could be looked for [NOW 10, BRI 11]. Using H_2 as fluidization gas to perform hydropyrolysis could solve some of these issues [LIN 14], but it would lead to safety problem and H_2 should be recycled.

At the particle and mesoscales, the *in situ* rheology has shown that lignin produces a sticky and soft material during pyrolysis, like "hot caramel." For this reason, lignin forms agglomerates in feeding screws and with the sand of the fluidized bed [NOW 10]. Moreover, the study on the hydrodeoxygenation (HDO) of real lignin vapors has highlighted the important loss in carbon in oligomers (analyzed by high-resolution mass spectrometry) and coke.

For all these reasons, we think that the fast pyrolysis of lignin in fluidized-bed reactors is not the most appropriate process for lignin conversion into BTX or monoaromatic phenols. The liquefaction of lignin in staged reactors (solvolysis followed by catalytic HDO) seems a better technology to produce aromatics, and our group is currently developing this process.

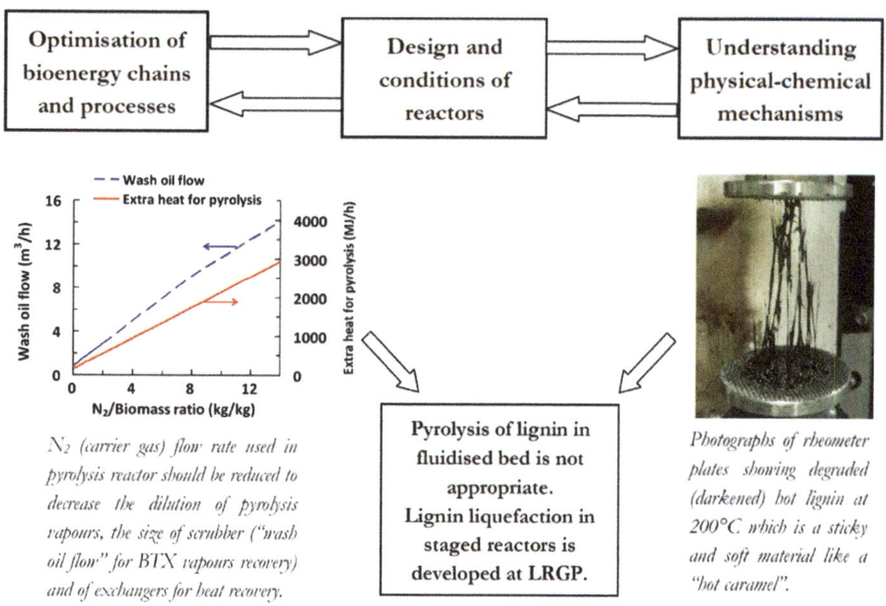

Figure 1. *Example of integrated multiscale approach on lignin valorization into aromatics*

General conclusion

In this book, we have shown the high potential of biomass valorization for the production of energy (fuels, electricity and heat) and chemicals (e.g. green aromatics).

Nevertheless, bioenergy chains can still be considerably improved.

The first major obstacle to tackle is policy or regulatory uncertainties, which does not provide long-term incentives to promote biomass. In this scope, the settlement of a global carbon economy may help the development of biomass.

From a technical point of view, the main improvement of processes could be:

– developing cleaner cookstove or combustion unit with lower pollutant emissions, limited ash agglomeration and higher energy efficiency;

– producing fast pyrolysis bio-oil of good quality (low particle content, low content in very reactive oxygenated moieties) at an industrial scale and with a tested commercial application (e.g. added-value chemicals, combustion or biofuels);

– improving the existing gasification technologies to make them more versatile and flexible to various biomasses and wastes (e.g. refuse-derived fuels), for instance in terms of bed agglomeration (by ashes), N or Cl species release;

– developing carbon sequestration technologies (through biochar or biogenic CO_2 capture by chemical looping) to get new routes for bioenergy supply with net negative emissions;

– developing staged, selective and robust catalytic systems able to convert light and heavy reactive species in bio-oils into biofuels or chemicals;

– revisiting old char production technologies and "intensifying" these processes with continuous and high-pressure reactors to improve char properties and yields;

– concerning liquefaction, recycling (green) solvent and a selective stable catalyst (or find a very cheap one), revisiting old coal liquefaction technologies, etc.

Numerous technologies have been developed in the past and they have been sometimes "forgotten" due to the cyclic research in bioenergy (cycles every ~20 years: the two World Wars, 70s to early 80s, nowadays, etc.).

From a more scientific perspective, which is obviously linked to the technical developments, future work could deal with:

– the development of multiscale modeling platform including process chains to molecular (chemical kinetics) aspects; such a platform may help in optimizing the chain based on physical fundamentals ("designing a biorefinery from Shrödinger equations"!);

– the development of in situ spectroscopic methods (and high-resolution imaging techniques) to monitor in real time, in the high-temperature (high-pressure) reactors, the thermal and chemical "history" of biomass particles, liquids or catalysts;

– the use of more theoretical studies (*ab initio*) for understanding the thermal or catalytic reactions;

– the use of more high-resolution mass spectrometry (petroleomic) methods to assess the selectivity of the reactors;

– the combination of thermochemical (fast but not selective) and biochemical (slow but selective) processes;

– and so on.

From an environmental point of view, the balance of nutrients and their return to the soil may be an important issue if more intensive scenarios for biomass production are favored. The emission of pollutants (such as black carbons from cookstove) is the major environmental and health impact of bioenergy, notably in the developing countries.

Biomass valorization units should be small to medium scale (decentralized) to promote local employments and to match with a sustainable production of the resource.

From a social point of view, the development of more efficient cookstoves, with a local value chain, will improve women conditions, employments, health and education of children. The critical elements to advancing sustained adoption of clean stoves are improving access to financing, user-centered design, filed testing, understanding cultural values and expectations and building local capacity [ANE 13].

Finally, the main limitation for biomass development is the price of fossil resource.

Bibliography

[ABD 12] ABDELOUAHED L., AUTHIER O., MAUVIEL G., *et al.*, "Detailed modeling of biomass gasification in dual fluidized bed reactors under Aspen Plus", *Energy & Fuels*, vol. 26, pp. 3840–3855, 2012.

[ABU 04] ABU EL-RUB Z., BRAMER E.A., BREM G., "Review of catalysts for tar elimination in biomass gasification processes", *Industrial & Engineering Chemistry Research*, vol. 43, pp. 6911–6919, 2004.

[AHO 10] AHO A., KUMAR N., LASHKUL A., *et al.*, "Catalytic upgrading of woody biomass derived pyrolysis vapours over iron modified zeolites in a dual-fluidized bed reactor", *Fuel*, vol. 89, pp. 1992–2000, 2010.

[ANE 13] ANENBERG S.C., BALAKRISHNAN K., JETTER J., et al., "Cleaner cooking solutions to achieve health, climate, and economic cobenefits", *Environmental Science & Technology*, vol. 47, pp. 3944–3952, 2013.

[ANA 03] ANTAL M.J., GRØNLI M., "The art, science, and technology of charcoal production", *Industrial & Engineering Chemistry Research*, vol. 42, pp. 1619–1640, 2003.

[AUT 15] AUTHIER O., Bioelectricity production, personal communication, EDF, 2015.

[AUT 09] AUTHIER O., FERRER M., MAUVIEL G., *et al.*, "Wood fast pyrolysis: comparison of Lagrangian and Eulerian modeling approaches with experimental measurements", *Industrial & Engineering Chemistry Research*, vol. 48, pp. 4796–4809, 2009.

[BAC 09] BACCILE N., LAURENT G., BABONNEAU F., *et al.*, "Structural characterization of hydrothermal carbon spheres by advanced solid-state MAS ^{13} C NMR investigations", *Journal of Physical Chemistry C*, vol. 113, pp. 9644–9654, 2009.

[BAD 11] BADAWI M., PAUL J.F., CRISTOL S., *et al.*, "Effect of water on the stability of Mo and CoMo hydrodeoxygenation catalysts: a combined experimental and DFT study", *Journal of Catalysis*, vol. 282, pp. 155–164, 2011.

[BAL 03] BALIGA V., SHARMA R., MISER D., *et al.*, "Physical characterization of pyrolyzed tobacco and tobacco components", *Journal of Analytical and Applied Pyrolysis*, vol. 66, pp. 191–215, 2003.

[BAR 97] BARDET M., EMSLEY L., VINCENDON M., "Two-dimensional spin-exchange solid-state NMR studies of 13C-enriched wood", *Solid State Nuclear Magnetic Resonance*, vol. 8, pp. 25–32, 1997.

[BAR 09] BARDET M., GERBAUD G., GIFFARD M., *et al.*, "13C high-resolution solid-state NMR for structural elucidation of archaeological woods", *Progress in Nuclear Magnetic Resonance Spectroscopy*, vol. 55, pp. 199–214, 2009.

[BAR 07] BARDET M., HEDIGER S., GERBAUD G., *et al.*, "Investigation with 13C NMR, EPR and magnetic susceptibility measurements of char residues obtained by pyrolysis of biomass", *Fuel*, vol. 86, pp. 1966–1976, 2007.

[BAU 05] BAUMLIN S., BROUST F., FERRER M., *et al.*, "The continuous self stirred tank reactor: measurement of the cracking kinetics of biomass pyrolysis vapours", *Chemical Engineering Science*, vol. 60, pp. 41–55, 2005.

[BRA 49] BRAQUE R., "Les industries de la carbonisation du bois en France", *Information Géographique*, vol. 13, pp. 28–33, 1949.

[BRI 12] BRIDGWATER A.V., "Review of fast pyrolysis of biomass and product upgrading", *Biomass and Bioenergy*, vol. 38, pp. 68–94, 2012.

[BRI 02] BRIDGWATER A.V., TOFT A.J., BRAMMER J.G., "A techno-economic comparison of power production by biomass fast pyrolysis with gasification and combustion", *Renewable and Sustainable Energy Reviews*, vol. 6, pp. 181–246, 2002.

[BRI 11] BRIENS C., "Lignin Pyrolysis in a Fluidized Bed Reactor with Fractional Condensation", in *the International Conference on Thermochemical Conversion Science*, Chicago, 2011.

[BRO 73] BROIDO A., JAVIER-SON A.C., OUANO A.C., *et al.*, "Molecular weight decrease in the early pyrolysis of crystalline and amorphous cellulose", *Journal of Applied Polymer Science*, vol. 17, pp. 3627–3635, 1973.

[BRO 10] BROSSE N., EL HAGE R., CHAOUCH M., *et al.*, "Investigation of the chemical modifications of beech wood lignin during heat treatment", *Polymer Degradation and Stability*, vol. 95, pp. 1721–1726, 2010.

[BRO 15] BROWN T.R., "A techno-economic review of thermochemical cellulosic biofuel pathways", *Bioresource Technology*, vol. 178, pp. 166–176, 2015.

[BUI 11] BUI V.N., LAURENTI D., DELICHÈRE P., *et al.*, "Hydrodeoxygenation of guaiacol", *Applied Catalysis B: Environmental*, vol. 101, pp. 246–255, 2011.

[BUL 11] BULUSHEV D.A., ROSS J.R.H., "Catalysis for conversion of biomass to fuels via pyrolysis and gasification: a review", *Catalysis Today*, vol. 171, pp. 1–13, 2011.

[CAO 12] CAO X., PIGNATELLO J.J., LI Y., *et al.*, "Characterization of wood chars produced at different temperatures using advanced solid-state 13C NMR spectroscopic techniques", *Energy & Fuels*, vol. 26, pp. 5983–5991, 2012.

[CAR 10] CARLSON T.R., CHENG Y.-T., JAE J., *et al.*, "Production of green aromatics and olefins by catalytic fast pyrolysis of wood sawdust", *Energy & Environmental Science*, vol. 4, pp. 145–161, 2010.

[CAR 07] CARPENTER D.L., DEUTCH S.P., FRENCH R.J., "Quantitative measurement of biomass gasifier tars using a molecular-beam mass spectrometer: comparison with traditional impinger sampling", *Energy & Fuels*, vol. 21, pp. 3036–3043, 2007.

[CEN 95] CENTENO A., LAURENT E., DELMON B., "Influence of the support of CoMo sulfide catalysts and of the addition of potassium and platinum on the catalytic performances for the hydrodeoxygenation of carbonyl, carboxyl, and guaiacol-type molecules", *Journal of Catalysis*, vol. 154, pp. 288–298, 1995.

[CHA 09] CHAIWAT W., HASEGAWA I., TANI T., *et al.*, "Analysis of cross-linking behavior during pyrolysis of cellulose for elucidating reaction pathway", *Energy & Fuels*, vol. 23, pp. 5765–5772, 2009.

[CHA 85] CHAN W.-C.R., KELBON M., KRIEGER B.B., "Modelling and experimental verification of physical and chemical processes during pyrolysis of a large biomass particle", *Fuel*, vol. 64, 1505–1513, 1985.

[CHE 09] CHERUBINI F., JUNGMEIER G., WELLISCH M., *et al.*, "Toward a common classification approach for biorefinery systems", *Biofuels Bioproducts and Biorefining*, vol. 3, pp. 534–546, 2009.

[CHU 13] CHU S., SUBRAHMANYAM A.V., HUBER G.W., "The pyrolysis chemistry of a β-O-4 type oligomeric lignin model compound", *Green Chemistry*, vol. 15, pp. 125–136, 2013.

[COL 14] COLLARD F.-X., BLIN J., "A review on pyrolysis of biomass constituents: Mechanisms and composition of the products obtained from the conversion of cellulose, hemicelluloses and lignin", *Renewable and Sustainable Energy Reviews*, vol. 38, pp. 594–608, 2014.

[CZE 04] CZERNIK S., BRIDGWATER A.V., "Overview of applications of biomass fast pyrolysis oil", *Energy & Fuels*, vol. 18, pp. 590–598, 2004.

[DAV 07] DAVIDIAN T., GUILHAUME N., IOJOIU E., *et al.*, "Hydrogen production from crude pyrolysis oil by a sequential catalytic process", *Applied Catalysis B: Environmental*, vol. 73, pp. 116–127, 2007.

[DEB 02] DE BRUIJN H., VAN DUIN R., HUIJBREGTS M.A.J., *Handbook on Life Cycle Assessment, Eco-Efficiency in Industry and Science*, Springer, Dordrecht, the Netherlands, 2002.

[DEL 08] DEGLISE X., "Trend and opportunities of bio-based industries," *FORTROPII*, Bangkok, Thailand, 2008.

[DEG 82] DEGLISE, X., "Les conversions thermochimiques du bois", *Revue Forestière Francaise*, 1982, vol. 4, pp. 249–271.

[DEG 91] DEGLISE X., MAGNE P., DONNOT A., *et al.*, "Catalytic pyrolysis," in GRASSI G., COLLINA A., ZIBETTA H. (eds), *Biomass for Energy, Industry, and Environment: Proceedings of the 6th International Conference on Biomass*, pp. 650–656, Elsevier Applied Science, London, 1991.

[DES 92] DESSUS B., DEVIN B., PHARABOD F., *Le potentiel mondial des énergies renouvelables*, Houille Blanche, France, 1992.

[DEW 11] DE WILD, P., IEA Bioenergy Agreement Task 34 Newsletter – PyNE 29, 2011.

[DEW 12] DE WILD P., HUIJGEN W.J.J., HEERES H.J., "Pyrolysis of wheat straw-derived organosolv lignin", *Journal of Analytical and Applied Pyrolysis*, vol. 93, pp. 95–103, 2012.

[DEW 09] DE WILD P., VAN DER LAAN R., KLOEKHORST A., *et al.*, "Lignin valorisation for chemicals and (transportation) fuels via (catalytic) pyrolysis and hydrodeoxygenation", *Environmental Progress & Sustainable Energy*, vol. 28, pp. 461–469, 2009.

[DÍA 08] DÍAZ M.C., EDECKI L., STEEL K.M., *et al.*, "Determination of the effects caused by different polymers on coal fluidity during carbonization using high-temperature ^{1}H NMR and rheometry", *Energy & Fuels*, vol. 22, pp. 471–479, 2008.

[DÍA 05] DÍAZ M.C., STEEL K.M., DRAGE T.C., *et al.*, "Determination of the effect of different additives in coking blends using a combination of in situ high-temperature ^{1}H NMR and rheometry", *Energy & Fuels*, vol. 19, pp. 2423–2431, 2005.

[DIF 11] DI FELICE L., COURSON C., FOSCOLO P.U., *et al.*, "Iron and nickel doped alkaline-earth catalysts for biomass gasification with simultaneous tar reformation and CO2 capture", *International Journal of Hydrogen Energy*, vol. 36, pp. 5296–5310, 2011.

[DON 08] DONOHOE B.S., DECKER S.R., TUCKER M.P., *et al.*, "Visualizing lignin coalescence and migration through maize cell walls following thermochemical pretreatment", *Biotechnology and Bioengineering*, vol. 101, pp. 913–925, 2008.

[DUF 04] DUFOUR, A., Potential of biomass for energy in France, Internal report for GDF, 2004.

[DUF 09a] DUFOUR A., GIRODS P., MASSON E., *et al.*, "Synthesis gas production by biomass pyrolysis: Effect of reactor temperature on product distribution", *International Journal of Hydrogen Energy*, vol. 34, pp. 1726–1734, 2009.

[DUF 09b] DUFOUR A., VALIN S., CASTELLI P., *et al.*, "Mechanisms and kinetics of methane thermal conversion in a syngas", *Industrial & Engineering Chemistry Research*, vol. 48, pp. 6564–6572, 2009.

[DUF 11a] DUFOUR A., MASSON E., GIRODS P., *et al.*, "Evolution of aromatic tar composition in relation to methane and ethylene from biomass pyrolysis-gasification", *Energy & Fuels*, vol. 25, pp. 4182–4189, 2011.

[DUF 11b] DUFOUR A., OUARTASSI B., BOUNACEUR R., *et al.*, "Modelling intra-particle phenomena of biomass pyrolysis", *Chemical Engineering Research and Design*, vol. 89, pp. 2136–2146, 2011.

[DUF 12a] DUFOUR A., CASTRO-ĎAZ M., MARCHAL P., *et al.*, "In situ analysis of biomass pyrolysis by high temperature rheology in relations with 1H NMR", *Energy & Fuels*, vol. 26, pp. 6432–6441, 2012.

[DUF 12b] DUFOUR A., CASTRO-DIAZ M., BROSSE N., *et al.*, "The origin of molecular mobility during biomass pyrolysis as revealed by in situ 1H NMR spectroscopy", *ChemSusChem*, vol. 5, pp. 1258–1265, 2012.

[DUF 12c] DUFOUR-KOWALSKI S., COURBAUD B., DREYFUS P., *et al.*, "Capsis: an open software framework and community for forest growth modelling", *Annals of Forest Science*, vol. 69, pp. 221–233, 2012.

[DUF 13a] DUFOUR A., "Geological sequestration of biomass char to mitigate climate change", *Environmental Science & Technology*, vol. 47, pp. 10106–10107, 2013.

[DUF 13b] DUFOUR A., WENG J., JIA L., *et al.*, "Revealing the chemistry of biomass pyrolysis by means of tunable synchrotron photoionisation-mass spectrometry", *RSC Advances*, vol. 3, pp. 4786–4792, 2013.

[DUF 15] DUFOUR A., CELZARD A., FIERRO V., *et al.*, "Catalytic conversion of methane over a biomass char for hydrogen production: deactivation and regeneration by steam gasification", *Applied Catalysis A: General*, vol. 490, pp. 170–180, 2015.

[EFF 08] EFFENDI A., GERHAUSER H., BRIDGWATER A.V., "Production of renewable phenolic resins by thermochemical conversion of biomass: a review", *Renewable and Sustainable Energy Reviews*, vol. 12, pp. 2092–2116, 2008.

[EHR 82] EHRBURGER P., LAHAYE J., WOZNIAK E., "Effect of carbonization on the porosity of beechwood", *Carbon*, vol. 20, pp. 433–439, 1982.

[EHR 89] EHRBURGER P., LOUYS F., LAHAYE J., "The concept of active sites applied to the study of carbon reactivity", *Carbon*, vol. 27, pp. 389–393, 1989.

[ELL 15a] ELLIOTT D.C., "Biofuel from fast pyrolysis and catalytic hydrodeoxygenation", *Current Opinion in Chemical Engineering*, vol. 9, pp. 59–65, 2015.

[ELL 91] ELLIOTT D.C., BECKMAN D., BRIDGWATER A.V., *et al.*, "Developments in direct thermochemical liquefaction of biomass: 1983-1990", *Energy & Fuels*, vol. 5, pp. 399–410, 1991.

[ELL 15b] ELLIOTT D.C., BILLER P., ROSS A.B., *et al.*, "Hydrothermal liquefaction of biomass: Developments from batch to continuous process", *Bioresource Technology*, vol. 178, pp. 147–156, 2015.

[EMM 43] EMMETT P.H., SKAU N., "The catalytic hydrogenation of benzene over metal catalysts", *Journal of the American Chemical Society*, vol. 65, pp. 1029–1035, 1943.

[ERD 14] ERDOCIA X., PRADO R., CORCUERA M.Á., *et al.*, "Effect of different organosolv treatments on the structure and properties of olive tree pruning lignin", *Journal of Industrial and Engineering Chemistry*, vol. 20, pp. 1103–1108, 2014.

[EVA 87] EVANS R.J., MILNE T.A., "Molecular characterization of the pyrolysis of biomass. 1. Fundamentals", *Energy & Fuels*, vol. 1, pp. 123–137, 1987.

[FAR 10] FARAVELLI T., FRASSOLDATI A., MIGLIAVACCA G., *et al.*, "Detailed kinetic modeling of the thermal degradation of lignins", *Biomass & Bioenergy*, vol. 34, pp. 290–301, 2010.

[FEL 88] FELDMANN H.F., PAISLEY M.A., APPELBAUM H.R., *et al.*, Conversion of forest residues to a methane-rich gas in a high-throughput gasifier, Report No. PNL-6570, Battelle, Columbus Division, OH (USA), 1988.

[FEN 12] FENDT A., STREIBEL T., SKLORZ M., *et al.*, "On-line process analysis of biomass flash pyrolysis gases enabled by soft photoionization mass spectrometry", *Energy & Fuels*, vol. 26, pp. 701–711, 2012.

[FIS 02] FISHER T., HAJALIGOL M., WAYMACK B., *et al.*, "Pyrolysis behavior and kinetics of biomass derived materials", *Journal of Analytical and Applied Pyrolysis*, vol. 62, pp. 331–349, 2002.

[FOR 12] FORTIN M., NINGRE F., ROBERT N., *et al.*, "Quantifying the impact of forest management on the carbon balance of the forest-wood product chain: a case study applied to even-aged oak stands in France", *Forest Ecology and Management*, vol. 279, pp. 176–188, 2012.

[FOS 14] FOSTON M., "Advances in solid-state NMR of cellulose", *Current Opinion in Biotechnology*, vol. 27, pp. 176–184, 2014.

[FOS 12] FOSTON M., SAMUEL R., RAGAUSKAS A.J., "13C cell wall enrichment and ionic liquid NMR analysis: progress towards a high-throughput detailed chemical analysis of the whole plant cell wall", *Analyst*, vol. 137, pp. 3904–3909, 2012.

[FRA 13a] FRANÇOIS J., ABDELOUAHED L., MAUVIEL G., *et al.*, "Detailed process modeling of a wood gasification combined heat and power plant", *Biomass & Bioenergy*, vol. 51, pp. 68–82, 2013.

[FRA 14] FRANÇOIS J., FORTIN M., PATISSON F., *et al.*, "Assessing the fate of nutrients and carbon in the bioenergy chain through the modeling of biomass growth and conversion", *Environmental Science & Technology*, vol. 48, pp. 14007–14015, 2014.

[FRA 13b] FRANÇOIS J., MAUVIEL G., FEIDT M., *et al.*, "Modeling of a biomass gasification CHP plant: influence of various parameters on energetic and exergetic efficiencies", *Energy & Fuels*, vol. 27, pp. 7398–7412, 2013.

[FRI 01] FRIDERICHSEN A.V., SHIN E.-J., EVANS R.J., *et al.*, "The pyrolysis of anisole (C6H5OCH3) using a hyperthermal nozzle", *Fuel*, vol. 80, pp. 1747–1755, 2001.

[GAS 11] GASTON K.R., JARVIS M.W., PEPIOT P., *et al.*, "Biomass pyrolysis and gasification of varying particle sizes in a fluidized-bed reactor", *Energy & Fuels*, vol. 25, pp. 3747–3757, 2011.

[GEL 73] GELDART D., "Types of gas fluidization", *Powder Technology*, vol. 7, pp. 285–292, 1973.

[GOH 66] GOHEEN D., "Hydrogenation of Lignin by the Noguchi Process," in MARTON J. (ed.), *Lignin Structure and Reactions*, Advances in Chemistry, Chapter 14, pp. 205–225, 1966.

[GOR 63] GORING D.A.I., "Thermal softening of lignin, hemicellulose, and cellulose", *Pulp & Paper Canada*, vol. 64, pp. T-517–T-527, 1963.

[GRA 84] GRAHAM R.G., BERGOUGNOU M.A., OVEREND R.P., "Fast pyrolysis of biomass", *Journal of Analytical and Applied Pyrolysis*, vol. 6, pp. 95–135, 1984.

[GUE 91] GUEDIRA F., BENNOUNA A., MAGNE P., MASSON D., "Etude comparative de la pyrolyse de déchets agro-alimentaires (tourteaux d'olives et bagasse) et de bois. Effet de la pression sur les rendements." *Bois et Fôrets des Tropiques*, 1991, no. 230, pp. 67–76.

[HAA 09] HAAS T.J., NIMLOS M.R., DONOHOE B.S., "Real-time and post-reaction microscopic structural analysis of biomass undergoing pyrolysis", *Energy & Fuels*, vol. 23, pp. 3810–3817, 2009.

[HAG 10] HAGE R.E., BROSSE N., SANNIGRAHI P., *et al.*, "Effects of process severity on the chemical structure of Miscanthus ethanol organosolv lignin", *Polymer Degradation and Stability*, vol. 95, pp. 997–1003, 2010.

[HAK 09] HAKKA M.H., GLAUDE P.-A., HERBINET O., *et al.*, "Experimental study of the oxidation of large surrogates for diesel and biodiesel fuels", *Combustion and Flame*, vol. 156, pp. 2129–2144, 2009.

[HAT 88] HATCHER P.G., "Dipolar-dephasing **1**3C NMR studies of decomposed wood and coalified xylem tissue: evidence for chemical structural changes associated with defunctionalization of lignin structural units during coalification", *Energy & Fuels*, vol. 2, pp. 48–58, 1988.

[HAU 14] HAUGEN N.E.L., TILGHMAN M.B., MITCHELL R.E., "The conversion mode of a porous carbon particle during oxidation and gasification", *Combustion and Flame*, vol. 161, pp. 612–619, 2014.

[HAV 08] HAVEREN J., SCOTT E.L., SANDERS J., "Bulk chemicals from biomass", *Biofuels, Bioproducts and Biorefining*, vol. 2, pp. 41–57, 2008.

[HAW 84] HAW J.F., MACIEL G.E., SCHROEDER H.A., "Carbon-13 nuclear magnetic resonance spectrometric study of wood and wood pulping with cross polarization and magic-angle spinning", *Analytical Chemistry*, vol. 56, pp. 1323–1329, 1984.

[HEN 09] HENRICH E., DAHMEN N., DINJUS E., "Cost estimate for biosynfuel production via biosyncrude gasification" *Biofuels, Bioproducts and Biorefining*, vol. 3, no. 1, January/February 2009, pp. 28–41.

[HOL 07] HOLLADAY J., BOZELL J., WHITE J., Top value-added chemicals from biomass volume II—results of screening for potential candidates from biorefinery lignin, Report No. PNNL-16983, U.S. Department of Energy, 2007.

[HOS 07] HOSOYA T., KAWAMOTO H., SAKA S., "Pyrolysis behaviors of wood and its constituent polymers at gasification temperature", *Journal of Analytical and Applied Pyrolysis*, vol. 78, pp. 328–336, 2007.

[HUB 08] HUBER G.W., Breaking the chemical & engineering barriers to lignocellulosic biofuels: next generation hydrocarbon biorefineries, NSF report, 2008.

[HUB 07] HUBER G.W., CORMA A., "Synergies between bio- and oil refineries for the production of fuels from biomass", *Angewandte Chemie International Edition in English*, vol. 46, pp. 7184–7201, 2007.

[HUB 06] HUBER G.W., IBORRA S., CORMA A., "Synthesis of transportation fuels from biomass: chemistry, catalysts, and engineering", *Chemical Reviews*, vol. 106, pp. 4044–4098, 2006.

[IEA 14] IEA, Key world energy statistics, available at: www.iea.org, 2014.

[IEA 12] IEA, Technology roadmap bioenergy for heat and power, available at: www.iea.org, 2012.

[INA 07] INARI G.N., MOUNGUENGUI S., DUMARÇAY S., *et al.*, "Evidence of char formation during wood heat treatment by mild pyrolysis", *Polymer Degradation and Stability*, vol. 92, pp. 997–1002, 2007.

[JAB 09] JABLONSKI W., GASTON K.R., NIMLOS M.R., *et al.*, "Pilot-scale gasification of corn stover, switchgrass, wheat straw, and wood: 2. Identification of global chemistry using multivariate curve resolution techniques", *Industrial & Engineering Chemistry Research*, vol. 48, pp. 10691–10701, 2009.

[JAE 11] JAE J., TOMPSETT G.A., FOSTER A.J., *et al.*, "Investigation into the shape selectivity of zeolite catalysts for biomass conversion", *Journal of Catalysis*, vol. 279, pp. 257–268, 2011.

[JAK 95] JAKAB E., FAIX O., TILL F., *et al.*, "Thermogravimetry/mass spectrometry study of six lignins within the scope of an international round robin test", *Journal of Analytical and Applied Pyrolysis*, vol. 35, pp. 167–179, 1995.

[JAR 00] JARVIS M.C., MCCANN M.C., "Macromolecular biophysics of the plant cell wall: concepts and methodology", *Plant Physiology and Biochemistry*, vol. 38, pp. 1–13, 2000.

[JES 96] JESS A., "Mechanisms and kinetics of thermal reactions of aromatic hydrocarbons from pyrolysis of solid fuels", *Fuel*, vol. 75, pp. 1441–1448, 1996.

[KAD 02] KADLA J.F., KUBO S., VENDITTI R.A., *et al.*, "Lignin-based carbon fibers for composite fiber applications", *Carbon*, vol. 40, pp. 2913–2920, 2002.

[KAW 07] KAWAMOTO H., HORIGOSHI S., SAKA S., "Pyrolysis reactions of various lignin model dimers", *Journal of Wood Science*, vol. 53, pp. 168–174, 2007.

[KER 13] KERSTEN S., GARCIA-PEREZ M., "Recent developments in fast pyrolysis of ligno-cellulosic materials", *Current Opinion in Biotechnology*, vol. 24, pp. 414–420, 2013.

[KIL 65] KILZER F.J., BROIDO A., "The nature of cellulose pyrolysis", *Pyrodynamics*, vol. 2, pp. 151–163, 1965.

[KIM 10] KIM H., RALPH J., "Solution-state 2D NMR of ball-milled plant cell wall gels in DMSO-d6/pyridine-d5", *Organic and Biomolecular Chemistry*, vol. 8, pp. 576–591, 2010.

[KIN 09] KINCHIN C.M., BAIN R.L., Hydrogen production from biomass via indirect gasification: the impact of NREL process development unit gasifier correlations, Technical Report NREL/TP, NREL, Golden, CO, 2009.

[KIM 11] KIRNBAUER F., HOFBAUER H., "Investigations on bed material changes in a dual fluidized bed steam gasification plant in Güssing, Austria", *Energy & Fuels*, vol. 25, pp. 3793–3798, 2011.

[KLA 10] KLASON P., HEIDENSTOM G.V., NORLIN E., "Untersuchungen zur Holzverkohlung" *Angewandte Chemie*, vol. 23, pp. 1252–1257, 1910.

[KLO 05] KLOSE W., WÖLKI M., "On the intrinsic reaction rate of biomass char gasification with carbon dioxide and steam", *Fuel*, vol. 84, pp. 885–892, 2005.

[KON 02] KONO H., YUNOKI S., SHIKANO T., *et al.*, "CP/MAS ^{13}C NMR study of cellulose and cellulose derivatives. 1. Complete assignment of the CP/MAS ^{13}C NMR spectrum of the native cellulose", *Journal of the American Chemical Society*, vol. 124, pp. 7506–7511, 2002.

[KRE 09] KREYSA G., "Sustainable management of the global carbon cycle through geostorage of wood", *ChemSusChem*, vol. 2, pp. 633–644, 2009.

[KSH 14] KSHIRSAGAR M.P., KALAMKAR, V.R., "A comprehensive review on biomass cookstoves and a systematic approach for modern cookstove design", *Renewable and Sustainable Energy Reviews*, vol. 30, pp. 580–603, 2014.

[KUM 14] KUMAR S., LANGE J.-P., VAN ROSSUM G., *et al.*, "Liquefaction of lignocellulose: process parameter study to minimize heavy ends", *Industrial & Engineering Chemistry Research*, vol. 53, pp. 11668–11676, 2014.

[LAN 63] LAINE N.R., VASTOLA F.J., WALKER P.L., "The importance of active surface area in the carbon-oxygen reaction", *Journal of Physical Chemistry*, vol. 67, pp. 2030–2034, 1963.

[LEB 15a] LE BRECH Y., Mechanisms of biomass primary pyrolysis, PhD Thesis, CNRS-University of Lorraine, 2015.

[LEB 15b] LE BRECH Y., DELMOTTE L., RAYA J., *et al.*, "High resolution solid state 2D NMR analysis of biomass and biochar", *Analytical Chemistry*, vol. 87, pp. 843–847, 2015.

[LÉD 02] LÉDÉ J., "Radiant flash pyrolysis of cellulose pellets: products and mechanisms involved in transient and steady state conditions", *Fuel*, vol. 81, pp. 1269–1279, 2002.

[LÉD 10] LÉDÉ J., "Biomass pyrolysis: comments on some sources of confusions in the definitions of temperatures and heating rates", *Energies*, vol. 3, pp. 886–898, 2010.

[LÉD 15] LÉDÉ J., AUTHIER O., "Temperature and heating rate of solid particles undergoing a thermal decomposition. Which criteria for characterizing fast pyrolysis?", *Journal of Analytical and Applied Pyrolysis*, vol. 113, pp. 1–14, 2015.

[LÉD 07] LÉDÉ J., BROUST F., NDIAYE F.-T., *et al.*, "Properties of bio-oils produced by biomass fast pyrolysis in a cyclone reactor", Fuel, vol. 86, pp. 1800–1810, 2007.

[LÉD 87] LÉDÉ J., LI H.Z., VILLERMAUX J., *et al.*, "Fusion-like behaviour of wood pyrolysis", *Journal of Analytical and Applied Pyrolysis*, vol. 10, pp. 291–308, 1987.

[LÉD 86] LÉDÉ J., VERZARO F., ANTOINE B., *et al.*, "Flash pyrolysis of wood in a cyclone reactor", *Chemical Engineering and Processing: Process Intensification*, vol. 20, pp. 309–317, 1986.

[LED 13] LEDESMA E.B., HOANG J.N., NGUYEN Q., *et al.*, "Unimolecular decomposition pathway for the vapor-phase cracking of eugenol, a biomass tar compound", *Energy & Fuels*, vol. 27, pp. 6839–6846, 2013.

[LEH 06] LEHMANN J., GAUNT J., RONDON M., "Bio-char sequestration in terrestrial ecosystems – a review", *Mitigation and Adaptation Strategies for Global Change*, vol. 11, pp. 395–419, 2006.

[LII 00] LIITIÄ T., MAUNU S.L., HORTLING B., "Solid-state NMR studies of residual lignin and its association with carbohydrates", *Journal of Pulp and Paper Science*, vol. 26, pp. 323–330, 2000.

[LIN 14] LINCK M., FELIX L., MARKER T., *et al.*, "Integrated biomass hydropyrolysis and hydrotreating: a brief review", *Wiley Interdisciplinary Reviews: Energy and Environment*, vol. 3, pp. 575–581, 2014.

[LIN 78] LINDEMUTH, T.E., "Investigations of the PERC process for biomass liquefaction at the Department of Energy, Albany, Oregon Experimental Facility", *Solid Wastes and Residues*, ACS Symposium Series, American Chemical Society, pp. 371–391, 1978.

[LOV 93] LOVE G.D., LAW R.V., SNAPE C.E., "Determination of nonprotonated aromatic carbon concentrations in coals by single pulse excitation 13C NMR", *Energy & Fuels*, vol. 7, pp. 639–644, 1993.

[MAM 09] MAMLEEV V., BOURBIGOT S., LE BRAS M., *et al.*, "The facts and hypotheses relating to the phenomenological model of cellulose pyrolysis", *Journal of Analytical and Applied Pyrolysis*, vol. 84, pp. 1–17, 2009.

[MAO 10] MAO J.D., HOLTMAN K.M., FRANQUI-VILLANUEVA D., "Chemical structures of corn stover and its residue after dilute acid prehydrolysis and enzymatic hydrolysis: insight into factors limiting enzymatic hydrolysis", *Journal of Agricultural and Food Chemistry*, vol. 58, pp. 11680–11687, 2010.

[MAO 12] MAO J.-D., JOHNSON R.L., LEHMANN J., *et al.*, "Abundant and stable char residues in soils: implications for soil fertility and carbon sequestration", *Environmental Science & Technology*, vol. 46, pp. 9571–9576, 2012.

[MAS 02] MASSIOT D., FAYON F., CAPRON M., *et al.*, "Modelling one- and two-dimensional solid-state NMR spectra", *Magnetic Resonance in Chemistry*, vol. 40, pp. 70–76, 2002.

[MAU 02] MAUNU S., "NMR studies of wood and wood products", *Progress in Nuclear Magnetic Resonance Spectroscopy*, vol. 40, pp. 151–174, 2002.

[MA 12] MA Z., TROUSSARD E., VAN BOKHOVEN J.A., "Controlling the selectivity to chemicals from lignin via catalytic fast pyrolysis", *Applied Catalysis A: General*, vol. 423–424, pp. 130–136, 2012.

[MEL 12] MELKIOR T., JACOB S., GERBAUD G., *et al.*, "NMR analysis of the transformation of wood constituents by torrefaction", *Fuel*, vol. 92, pp. 271–280, 2012.

[MIH 11] MIHALCIK D.J., BOATENG A.A., MULLEN C.A., *et al.*, "Packed-bed catalytic cracking of oak derived pyrolytic vapors", *Industrial & Engineering Chemistry Research*, vol. 50, pp. 13304–13312, 2011.

[MIL 96] MILOSAVLJEVIC I., OJA V., SUUBERG E.M., "Thermal effects in cellulose pyrolysis: relationship to char formation processes", *Industrial & Engineering Chemistry Research*, vol. 35, pp. 653–662, 1996.

[MIY 79] MIYAZAWA K., YOKONO T., SANADA Y., "High temperature 1H-NMR study of coal and pitch at the early stages of carbonization", *Carbon*, vol. 17, pp. 223–225, 1979.

[MOG 06] MOGHTADERI B., "The state-of-the-art in pyrolysis modelling of lignocellulosic solid fuels", *Fire and Materials*, vol. 30, pp. 1–34, 2006.

[MOK 83] MOK W.S.-L., ANTAL JR. M.J., "Effects of pressure on biomass pyrolysis. II. Heats of reaction of cellulose pyrolysis", *Thermochimica Acta*, vol. 68, pp. 165–186, 1983.

[MOL 08] MOLLERAT J.B., Mémoires présentés à l'institut sur la distillation du bois et l'emploi de ses produits, Bibliothèque Nationale de France, Paris, 1808.

[MOR 00] MORTEN G. GRØNLI, MORTEN C. MELAAN, "Mathematical Model for Wood Pyrolysis Comparison of Experimental Measurements with Model Predictions", *Energy Fuels*, vol. 14, no. 4, pp 791–800, 2000

[NAK 08] NAKAMURA T., KAWAMOTO H., SAKA S., "Pyrolysis behavior of Japanese cedar wood lignin studied with various model dimers", *Journal of Analytical and Applied Pyrolysis*, vol. 81, pp. 173–182, 2008.

[NAR 96] NARAYAN R., ANTAL M.J., "Thermal lag, fusion, and the compensation effect during biomass pyrolysis", *Industrial & Engineering Chemistry Research*, vol. 35, pp. 1711–1721, 1996.

[NEW 92] NEWMAN R.H., "Nuclear magnetic resonance study of spatial relationships between chemical components in wood cell walls", *Holzforschung*, pp. 205–201, 1992.

[NOW 14] NOWAKOWSKA M., HERBINET O., DUFOUR A., *et al.*, "Detailed kinetic study of anisole pyrolysis and oxidation to understand tar formation during biomass combustion and gasification", *Combustion and Flame*, vol. 161, pp. 1474–1488, 2014.

[NOW 10] NOWAKOWSKI D.J., BRIDGWATER A.V., ELLIOTT D.C., *et al.*, "Lignin fast pyrolysis: results from an international collaboration", *Journal of Analytical and Applied Pyrolysis*, vol. 88, pp. 53–72, 2010.

[NUS 03] NUSSBAUMER T., "Combustion and co-combustion of biomass: fundamentals, technologies, and primary measures for emission reduction", *Energy & Fuels*, vol. 17, pp. 1510–1521, 2003.

[OAS 03] OASMAA A., KUOPPALA E., "Fast pyrolysis of forestry residue. 3. Storage stability of liquid fuel", *Energy & Fuels*, vol. 17, pp. 1075–1084, 2003.

[OH 89] OH M.S., PETERS W.A., HOWARD J.B., "An experimental and modeling study of softening coal pyrolysis", *AIChE Journal*, vol. 35, pp. 775–792, 1989.

[OHN 77] OHNISHI A., KATŌ K., TAKAGI E., "Pyrolytic formation of 3-hydroxy-2-penteno-1,5-lactone from xylan, xylo-oligosaccharides, and methyl xylopyranosides", *Carbohydrate Research*, vol. 58, pp. 387–395, 1977.

[OLC 13a] OLCESE R., BETTAHAR M.M., MALAMAN B., *et al.*, "Gas-phase hydrodeoxygenation of guaiacol over iron-based catalysts. Effect of gases composition, iron load and supports (silica and activated carbon)", *Applied Catalysis B: Environmental*, vol. 129, pp. 528–538, 2013.

[OLC 13b] OLCESE R., CARRÉ V., AUBRIET F., *et al.*, "Selectivity of bio-oils catalytic hydrotreatment assessed by petroleomic and GC*GC/MS-FID analysis", *Energy & Fuels*, vol. 27, pp. 2135–2145, 2013.

[OLC 12] OLCESE R.N., BETTAHAR M., PETITJEAN D., *et al.*, "Gas-phase hydrodeoxygenation of guaiacol over Fe/SiO2 catalyst", *Applied Catalysis B: Environmental*, vol. 115–116, pp. 63–73, 2012.

[OLC 13c] OLCESE R.N., FRANCOIS J., BETTAHAR M.M., *et al.*, "Hydrodeoxygenation of guaiacol, a surrogate of lignin pyrolysis vapors, over iron based catalysts: kinetics and modeling of the lignin to aromatics integrated process", *Energy & Fuels*, vol. 27, pp. 975–984, 2013.

[OLC 13d] OLCESE R.N., LARDIER G., BETTAHAR M., *et al.*, "Aromatic chemicals by iron-catalyzed hydrotreatment of lignin pyrolysis vapor", *ChemSusChem*, vol. 6, pp. 1490–1499, 2013.

[PAN 11] PANDEY M.P., KIM C.S., "Lignin depolymerization and conversion: a review of thermochemical methods", *Chemical Engineering and Technology*, vol. 34, pp. 29–41, 2011.

[PAR 04] PARIKKA M., "Global biomass fuel resources", *Biomass & Bioenergy*, vol. 27, pp. 613–620, 2004.

[PEC 15] PECHA B., ARAUZO P., GARCIA-PEREZ M., "Impact of combined acid washing and acid impregnation on the pyrolysis of Douglas fir wood", *Journal of Analytical and Applied Pyrolysis*, vol. 114, pp. 127–137, 2015.

[PEL 15] PELLETIER C., Assessment of heat production from wood, PhD Thesis, CNRS-University of Lorraine, 2015.

[PEN 10] PENG Y., WU S., "The structural and thermal characteristics of wheat straw hemicellulose", *Journal of Analytical and Applied Pyrolysis*, vol. 88, pp. 134–139, 2010.

[PER 09] PERRÉ P., "Production de charbon de bois", *Conversion thermochimique de la biomasse*, Nancy, 2009.

[PFE 04] PFEIFER C., RAUCH R., HOFBAUER H., "In-bed catalytic tar reduction in a dual fluidized bed biomass steam gasifier", *Industrial & Engineering Chemistry Research*, vol. 43, pp. 1634–1640, 2004.

[PIS 86] PISKORZ J., RADLEIN D., SCOTT D.S., "On the mechanism of the rapid pyrolysis of cellulose", *Journal of Analytical and Applied Pyrolysis*, vol. 9, pp. 121–137, 1986.

[PIS 89] PISKORZ J., RADLEIN D.S.A.G., SCOTT D.S., *et al.*, "Pretreatment of wood and cellulose for production of sugars by fast pyrolysis", *Journal of Analytical and Applied Pyrolysis*, vol. 16, pp. 127–142, 1989.

[PON 91] PONDER G.R., RICHARDS G.N., "Thermal synthesis and pyrolysis of a xylan", *Carbohydrate Research*, vol. 218, pp. 143–155, 1991.

[POP 10] POPOV A., KONDRATIEVA E., GOUPIL J.M., *et al.*, "Bio-oils hydrodeoxygenation: adsorption of phenolic molecules on oxidic catalyst supports", *Journal of Physical Chemistry C*, vol. 114, pp. 15661–15670, 2010.

[POT 06] POTTING J., HERTEL O., SCHÖPP W., *et al.*, "Spatial differentiation in the characterisation of photochemical ozone formation: the EDIP2003 methodology", *International Journal of Life Cycle Assessment*, vol. 11, pp. 72–80, 2006.

[PRÉ 15] PRÉ CONSULTANTS, ReCiPe method, accessed 2015, Available at: www.pre-sustainability.com.

[RAD 91] RADOVIĆ L.R., LIZZIO A.A., JIANG H., "Reactive Surface Area: An Old But New Concept in Carbon Gasification", in LAHAYE J., EHRBURGER P. (eds), *Fundamental Issues in Control of Carbon Gasification Reactivity*, NATO ASI Series, Springer, the Netherlands, pp. 235–255, 1991.

[RAD 83] RADOVIĆ L.R., WALKER JR. P.L., JENKINS R.G., "Importance of carbon active sites in the gasification of coal chars", *Fuel*, vol. 62, pp. 849–856, 1983.

[RAN 08] RANZI E., CUOCI A., FARAVELLI T., *et al.*, "Chemical kinetics of biomass pyrolysis", *Energy & Fuels*, vol. 22, pp. 4292–4300, 2008.

[RAT 03] RATH J., WOLFINGER M.G., STEINER G., *et al.*, "Heat of wood pyrolysis", *Fuel*, vol. 82, pp. 81–91, 2003.

[RAV 96] RAVEENDRAN K., "Pyrolysis characteristics of biomass and biomass components", *Fuel*, vol. 75, pp. 987–998, 1996.

[REP 10] REPELLIN V., GOVIN A., ROLLAND M., *et al.*, "Energy requirement for fine grinding of torrefied wood", *Biomass & Bioenergy*, vol. 34, pp. 923–930, 2010.

[RIC 87] RICHARDS G.N., "Glycolaldehyde from pyrolysis of cellulose", *Journal of Analytical and Applied Pyrolysis*, vol. 10, pp. 251–255, 1987.

[RIE 11] RIEBEEK H., The carbon cycle, NASA, available at: http://earthobservatory.nasa.gov/Features/CarbonCycle/, 2011.

[ROD 93] RODRIGUEZ-MIRASOL J., CORDERO T., RODRIGUEZ J.J., "Activated carbons from carbon dioxide partial gasification of eucalyptus kraft lignin", *Energy & Fuels*, vol. 7, pp. 133–138, 1993.

[RØD 12] RØDSRUD G., LERSCH M., SJÖDE A., "History and future of world's most advanced biorefinery in operation", *Biomass & Bioenergy*, vo. 46, pp. 46–59, 2012.

[ROG 15] ROGAUME Y., Emission of wood combustion technologies, Personal communication, 2015.

[ROM 10] ROMERO Y., RICHARD F., BRUNET S., "Hydrodeoxygenation of 2-ethylphenol as a model compound of bio-crude over sulfided Mo-based catalysts: promoting effect and reaction mechanism", *Applied Catalysis B: Environmental*, vol. 98, pp. 213–223, 2010.

[ROU 90] ROUZAUD J.-N., "Contribution of transmission electron microscopy to the study of the coal carbonization processes", *Fuel Processing Technology*, vol. 24, pp. 55–69, 1990.

[SAK 93] SAKUROVS R., LYNCH L.J., "Direct observations on the interaction of coals with pitches and organic compounds during co-pyrolysis", *Fuel*, vol. 72, pp. 743–749, 1993.

[SAK 87] SAKUROVS R., LYNCH L.J., MAHER T.P., *et al.*, "Molecular mobility during pyrolysis of Australian bituminous coals", *Energy & Fuels*, vol. 1, pp. 167–172, 1987.

[SAL 09] SALMON E., BEHAR F., LORANT F., *et al.*, "Early maturation processes in coal. Part 1: pyrolysis mass balance and structural evolution of coalified wood from the Morwell Brown Coal seam", *Organic Geochemistry*, vol. 40, pp. 500–509, 2009.

[SAM 11] SAMUEL R., FOSTON M., JAING N., *et al.*, "HSQC (heteronuclear single quantum coherence) 13C- 1H correlation spectra of whole biomass in perdeuterated pyridinium chloride-DMSO system: an effective tool for evaluating pretreatment", *Fuel*, vol. 90, pp. 2836–2842, 2011.

[SCH 11] SCHEER A.M., MUKARAKATE C., ROBICHAUD D.J., *et al.*, "Thermal decomposition mechanisms of the methoxyphenols: formation of phenol, cyclopentadienone, vinylacetylene, and acetylene", *Journal of Physical Chemistry A*, vol. 115, pp. 13381–13389, 2011.

[SCH 01] SCHEIRS J., CAMINO G., TUMIATTI W., "Overview of water evolution during the thermal degradation of cellulose", *European Polymer Journal*, vol. 37, pp. 933–942, 2001.

[SHA 03] SHABTAI J.S., ZMIERCZAK W.W., CHORNET E., *et al.*, Process for converting lignins into a high octane blending component, Patent US Patent 20,030,115,792, 2003.

[SHA 04] SHARMA R.K., WOOTEN J.B., BALIGA V.L., *et al.*, "Characterization of chars from pyrolysis of lignin", *Fuel*, vol. 83, pp. 1469–1482, 2004.

[SHE 09] SHEN D.K., GU S., "The mechanism for thermal decomposition of cellulose and its main products", *Bioresource Technology*, vol. 100, pp. 6496–6504, 2009.

[SHE 10a] SHEN D.K., GU S., BRIDGWATER A.V., "Study on the pyrolytic behaviour of xylan-based hemicellulose using TG–FTIR and Py–GC–FTIR", *Journal of Analytical and Applied Pyrolysis*, vol. 87, pp. 199–206, 2010.

[SHE 10b] SHEN D.K., GU S., LUO K.H., *et al.*, "The pyrolytic degradation of wood-derived lignin from pulping process", *Bioresource Technology*, vol. 101, pp. 6136–6146, 2010.

[SHI 00] SHIN E.-J., KEANE M.A., "Gas-phase hydrogenation/hydrogenolysis of phenol over supported nickel catalysts", *Industrial & Engineering Chemistry Research*, vol. 39, pp. 883–892, 2000.

[SHI 01] SHIN E.-J., NIMLOS M.R., EVANS R.J., "Kinetic analysis of the gas-phase pyrolysis of carbohydrates", *Fuel*, vol. 80, pp. 1697–1709, 2001.

[SIV 02] SIVONEN H., MAUNU S.L., SUNDHOLM F., *et al.*, "Magnetic resonance studies of thermally modified wood", *Holzforschung*, vol. 56, pp. 648–654, 2002.

[SPA 05] SPATH P., ADEN A., EGGEMAN T., *et al.*, Biomass to hydrogen production detailed design and economics utilizing the Battelle Columbus Laboratory indirectly heated gasifier, Technical Report NREL/TP-510-37408 NREL/TP, NREL, Golden, CO, 2005.

[STE 04] STEEL K.M., DIAZ M.C., PATRICK J.W., *et al.*, "Use of rheometry and ^{1}H NMR spectroscopy for understanding the mechanisms behind the generation of coking pressure", *Energy & Fuels*, vol. 18, pp. 1250–1256, 2004.

[STO 90] STOECKLI H.F., "Microporous carbons and their characterization: the present state of the art", *Carbon*, vol. 28, pp. 1–6, 1990.

[SUT 01] SUTTON D., KELLEHER B., ROSS J.R.H., "Review of literature on catalysts for biomass gasification", *Fuel Processing Technology*, vol. 73, pp. 155–173, 2001.

[SUU 96] SUUBERG E.M., MILOSAVLJEVIC I., OJA V., "Two-regime global kinetics of cellulose pyrolysis: the role of tar evaporation", *Symposium (International) on Combustion*, vol. 26, pp. 1515–1521, 1996.

[TEK 87] TEKELY P., NICOLE D., DELPUECH J.-j., *et al.*, "Chemical structure changes in coals after low-temperature oxidation and demineralization by acid treatment as revealed by high resolution solid state 13C NMR", *Fuel Processing Technology*, vol. 15, pp. 225–231, 1987.

[VAL 12] VALLE B., CASTAÑO P., OLAZAR M., *et al.*, "Deactivating species in the transformation of crude bio-oil with methanol into hydrocarbons on a HZSM-5 catalyst", *Journal of Catalysis*, vol. 285, pp. 304–314, 2012.

[VAL 10] VALLE B., GAYUBO A.G., AGUAYO A.T., *et al.*, "Selective production of aromatics by crude bio-oil valorization with a nickel-modified HZSM-5 zeolite catalyst", *Energy & Fuels*, vol. 24, pp. 2060–2070, 2010.

[VAN 04] VAN DER DRIFT A., Entrained flow gasification of biomass ash behaviour, feeding issues, and system analyses, ECN report, 2004.

[VAN 13] VAN VUUREN D.P., DEETMAN S., VAN VLIET J., *et al.*, "The role of negative CO2 emissions for reaching 2 °C – insights from integrated assessment modelling", *Climatic Change*, vol. 118, pp. 15–27, 2013.

[VAS 09] VASILIOU A., NIMLOS M.R., DAILY J.W., *et al.*, "Thermal decomposition of furan generates propargyl radicals", *Journal of Physical Chemistry A*, vol. 113, pp. 8540–8547, 2009.

[VEN 10] VENDERBOSCH R., PRINS W., "Fast pyrolysis technology development", *Biofuels, Bioproducts and Biorefining*, vol. 4, pp. 178–208, 2010.

[VIL 09] VILLAVERDE J.J., LI J., EK M., *et al.*, "Native lignin structure of Miscanthus x giganteus and its changes during acetic and formic acid fractionation", *Journal of Agricultural and Food Chemistry*, vol. 57, pp. 6262–6270, 2009.

[VIR 12] VIRGINIE M., ADÁNEZ J., COURSON C., *et al.*, "Effect of Fe–olivine on the tar content during biomass gasification in a dual fluidized bed", *Applied Catalysis B: Environmental*, vol. 121–122, pp. 214–222, 2012.

[WAN 97] WANG H., FRENKLACH M., "A detailed kinetic modeling study of aromatics formation in laminar premixed acetylene and ethylene flames", *Combustion and Flame*, vol. 110, pp. 173–221, 1997.

[WAN 90] WANG N., LOW M.J.D., "Spectroscopic studies of carbons. XV. The pyrolysis of a lignin", *Materials Chemistry and Physics*, vol. 26, pp. 67–80, 1990.

[WAN 09] WANG S., WANG K., LIU Q., *et al.*, "Comparison of the pyrolysis behavior of lignins from different tree species", *Biotechnology Advances*, vol. 27, pp. 562–567, 2009.

[WEN 13] WEN J.-L., SUN S.-L., XUE B.-L., *et al.*, "Recent advances in characterization of lignin polymer by solution-state nuclear magnetic resonance (NMR) methodology", *Materials*, vol. 6, pp. 359–391, 2013.

[WER 14] WERNER K., POMMER L., BROSTRÖM M., "Thermal decomposition of hemicelluloses", *Journal of Analytical and Applied Pyrolysis*, vol. 110, pp. 130–137, 2014.

[WER 04] WERPY T., PETERSEN G., Top value added chemicals from biomass: Volume I – results of screening for potential candidates from sugars and synthesis gas, Report no. DOE/GO-102004-1992, National Renewable Energy Lab., Golden, CO, 2004.

[WOO 04] WOOTEN J.B., SEEMAN J.I., HAJALIGOL M.R., "Observation and characterization of cellulose pyrolysis intermediates by ^{13}C CPMAS NMR. A new mechanistic model", *Energy & Fuels*, vol. 18, pp. 1–15, 2004.

[WOR 07] WORASUWANNARAK N., SONOBE T., TANTHAPANICHAKOON W., "Pyrolysis behaviors of rice straw, rice husk, and corncob by TG-MS technique", *Journal of Analytical and Applied Pyrolysis*, vol. 78, pp. 265–271, 2007.

[YOO 83] YOON K.J., VANNICE M.A., "Benzene hydrogenation over iron: II. Reaction model over unsupported and supported catalysts", *Journal of Catalysis*, vol. 82, pp. 457–468, 1983.

[YUN 09] YUNG M.M., JABLONSKI W.S., MAGRINI-BAIR K.A., "Review of catalytic conditioning of biomass-derived syngas", *Energy & Fuels*, vol. 23, pp. 1874–1887, 2009.

[ZAC 14] ZACHER A.H., OLARTE M.V., SANTOSA D.M., *et al.*, "A review and perspective of recent bio-oil hydrotreating research", *Green Chemistry*, vol. 16, pp. 491–515, 2014.

[ZAK 10] ZAKZESKI J., BRUIJNINCX P.C.A., JONGERIUS A.L., *et al.*, "The catalytic valorization of lignin for the production of renewable chemicals", *Chemical Reviews*, vol. 110, pp. 3552–3599, 2010.

[ZAW 02] ZAWADZKI J., WISNIEWSKI M., "13C NMR study of cellulose thermal treatment", *Journal of Analytical and Applied Pyrolysis*, vol. 62, pp. 111–121, 2002.

[ZHE 08] ZENG N., "Carbon sequestration via wood burial", *Carbon Balance and Management*, vol. 3, pp. 1–12, 2008.

[ZHA 13] ZHANG J., WU R., ZHANG G., *et al.*, "Technical review on thermochemical conversion based on decoupling for solid carbonaceous fuels", *Energy & Fuels*, vol. 27, pp. 1951–1966, 2013.

[ZHA 11] ZHAO H.Y., LI D., BUI P., *et al.*, "Hydrodeoxygenation of guaiacol as model compound for pyrolysis oil on transition metal phosphide hydroprocessing catalysts", *Applied Catalysis A: General*, vol. 391, pp. 305–310, 2011.

[ZHO 14] ZHOU S., GARCIA-PEREZ M., PECHA B., *et al.*, "Effect of particle size on the composition of lignin derived oligomers obtained by fast pyrolysis of beech wood", *Fuel*, vol. 125, pp. 15–19, 2014.

[ZHU 11] ZHU X., LOBBAN L.L., MALLINSON R.G., *et al.*, "Bifunctional transalkylation and hydrodeoxygenation of anisole over a Pt/HBeta catalyst", *Journal of Catalysis*, vol. 281, pp. 21–29, 2011.

[ZOU 15] ZOULALIAN A., BOUNACEUR R., DUFOUR A., "Kinetic modelling of char gasification by accounting for the evolution of the reactive surface area", *Chemical Engineering Science*, vol. 138, pp. 281–290, 2015.

Index

M, N

O, P

S, T

V, X

Other titles from

ISTE

in

Energy

2015

CROGUENNEC Laurence, MONCONDUIT Laure, DEDRYVERE Rémi
Electrodes for Li-ion Batteries

ROBYNS Benoît, FRANÇOIS Bruno, DELILLE Gauthier, SAUDEMONT Christophe
Energy Storage in Electric Power Grids

ROSSI Carole
Al-based Energetic Nanomaterials

TARASCON Jean-Marie, SIMON Patrice
Electrochemical Energy Storage

2013

LALOUI Lyesse and DI DONNA Alice
Energy Geostructures: Innovation in Underground Engineering

2012

BECKERS Benoit
Solar Energy at Urban Scale

ROBYNS Benoît, DAVIGNY Arnaud, FRANÇOIS Bruno, HENNETON Antoine, SPROOTEN Jonathan
Electricity Production from Renewable Energies

2011

GAO Fei, BLUNIER Benjamin, MIRAOUI Abdellatif
Proton Exchange Membrane Fuel Cell Modeling

MULTON Bernard
Marine Renewable Energy Handbook

2010

BRUNET Yves
Energy Storage

2009

SABONNADIERE Jean-Claude
Low Emission Power Generation Technologies and Energy Management

SABONNADIERE Jean-Claude
Renewable Energy Technologies